# THE VULNERABLE PLANET

## A Short Economic History of the Environment

JOHN BELLAMY FOSTER

MONTHLY REVIEW PRESS
NEW YORK

*Library of Congress Cataloging-in-Publication Data*
Foster, John Bellamy.
The vulnerable planet: a short economic history
of the environment / by John Bellamy Foster. — [Rev. ed.]
p. cm.
Includes bibliographical references and index.
ISBN 1-58367-019-X (paper)
1. Environmental degradation. 2. Environmental policy. I. Title.

GE140.F68 1999
363.7—dc21 99-38834
CIP

Monthly Review Press
122 West 27th Street
New York, NY 10001

Manufactured in Canada
10 9 8 7 6

# CONTENTS

# PREFACE

I was slow in committing myself fully to the environmental cause. I grew up in the 1950s and 1960s in the Pacific Northwest of the United States, an area known for the quality of its environment. During the first Earth Day in April 1970 I was a fairly complacent participant. It was natural, I thought, that those living in L.A. or along the Great Lakes should be disturbed by the destruction of the environment. But in the Northwest we were as yet comparatively free from such worries. My chief concern at that time was the Vietnam War. As long as napalm was being used on the people of Indochina, the issue of the health of the environment seemed an unaffordable luxury.

From the mid-1970s to the mid-1980s the most important questions seemed to me to be those of economic crisis and third world underdevelopment. As a public intellectual in the Reagan era, my efforts were devoted primarily to resisting the attempt by the powers that be to shift the burden of economic stagnation onto the backs of workers, the unemployed, women, people of color, and the majority of the world's population in the third world. For me, as for Raymond Williams, "resistance to capitalism" had become the "the decisive form of the necessary human defense." But unlike Williams I did not yet fully appreciate the connection between this and the fate of the earth itself.[1]

It was my return to the Pacific Northwest in 1985, after almost a decade's absence, that finally forced me to open my eyes to the full global dimensions of the ecological crisis. "Every middle-aged man who revisits his birthplace after a few years of absence," New England environmentalist George Perkins Marsh observed in 1847, "looks upon another landscape." Marsh, then in his mid-forties, was alarmed by the cycles of destruction that he had witnessed over his own lifetime in his native Vermont. The devastation of hillsides, the burning of the woods, the damming of streams represented changes "too striking," he wrote, "to have escaped the attention of any observing person."[2]

In returning to the Northwest in the mid-1980s, I discovered that a region that had been known for the pristine character of much of its environment was now facing the prospect of massive, irreversible ecological destruction—of global, not simply regional, proportions. The old-growth coniferous forest, containing many of the world's oldest and tallest trees, was being cut down at a record rate; the Columbia–Snake River system had become one of the most threatened river systems in the country; and numerous species, from the northern spotted owl to the Snake River salmon, were endangered. The Hanford nuclear facility in the state of Washington was revealed to be one of the world's worst sources of radioactive contamination. Everywhere a war was raging between environmentalists and the forces of economic expansion.

The late 1980s were also years in which world awareness of the threats posed by global warming, destruction of the ozone layer, tropical deforestation, and the extinction of species rose dramat-

**Untreatable oil waste stored in barrels at an oil-waste treatment center in Bahia, Brazil. [Katherine McGlynn/Impact Visuals]**

ically. The combined effect of these influences compelled me to face up to the growing dangers to the planet.

Yet for all of this new ecological awareness, the social concerns of my youth were by no means lessened. There is a very real sense in which those who have devoted their energies to overcoming social inequality have always been involved in the battle for environmental justice. War, economic inequality, and third world underdevelopment—the three phenomena that drew most of my attention in the decades of the 1970s and 1980s—are inextricably bound to the larger question of the systematic degradation of the planet and of the conditions of life for a majority of the world's people. The failure to perceive this connection is the principal weakness of mainstream environmentalism. If we are to make sense of the environmental crisis, we must not abandon radical social concerns but broaden and deepen them to take into account the destruction of the earth itself. It is above all the *interconnectedness* of social and environmental problems that constitutes the primary basis upon which a potent movement for change can be created.

My intellectual and personal debts in the writing of this book are many. Among those who have directly or indirectly given support to this work, I would like to mention Michael Dawson, Susan Lowes, Paul Sweezy, Harry Magdoff, James O'Connor, Art MacEwan, Bill Tabb, Mike Tanzer, Bob McChesney, Patrick Novotny, David Milton, Nancy Milton, Ellen Meiksins Wood, Neal Wood, Edward Goldsmith, Greg McLauchlan, Sandy Ashton, Dave Ashton, Bill Foster, Linda Berentsen, Eileen Merritt, Charles Hunt, Kathy Hunt, and Laura Tamkin. *The Vulnerable Planet* itself is dedicated to Saul, whose all-embracing love of the earth and of humanity in the third year of his life has helped sustain within me an ecological politics of hope without which this work would not have been possible.

# 1

# THE ECOLOGICAL CRISIS

Human society has reached a critical threshold in its relation to its environment. The destruction of the planet, in the sense of making it unusable for human purposes, has grown to such an extent that it now threatens the continuation of much of nature, as well as the survival and development of society itself. The litany of ecological complaints plaguing the world today encompasses a long list of urgent problems. These include: overpopulation, destruction of the ozone layer, global warming, extinction of species, loss of genetic diversity, acid rain, nuclear contamination, tropical deforestation, the elimination of climax forests, wetland destruction, soil erosion, desertification, floods, famine, the despoliation of lakes, streams, and rivers, the drawing down and contamination of ground water, the pollution of coastal waters and estuaries, the destruction of coral reefs, oil spills, overfishing, expanding landfills, toxic wastes, the poisonous effects of insecticides and herbicides, exposure to hazards on the job, urban congestion, and the depletion of nonrenewable resources. According to the prestigious Worldwatch Institute, we have only four decades left in which to gain control over our major environmental problems if we are to avoid irreversible socio-ecological decline, and the 1990s are the critical decade in which the necessary changes must begin to occur.[1]

Yet most current prescriptions for solving the planet's ecological problems are woefully inadequate to meet such ominous threats, since they amount to little more than calls for new international agreements, for personal restraint with regard to the growth of both population and consumption, and the adoption of a handful of so-called environmentally friendly technologies. In what follows I will argue that we must begin by recognizing that the crisis of the earth is not a crisis of *nature* but a crisis of *society*. The chief causes of the environmental destruction that faces us today are not biological, or the product of individual human choice. They are social and historical, rooted in the productive relations, technological imperatives, and historically conditioned demographic trends that characterize the dominant social system. Hence, what is ignored or downplayed in most proposals to remedy the environmental crisis is the most critical challenge of all: the need to transform the major social bases of environmental degradation, and not simply to tinker with its minor technical bases. As long as prevailing social relations remain unquestioned, those who are concerned about what is happening are left with few visible avenues for environmental action other than purely personal commitments to recycling and green shopping, socially untenable choices between jobs and the environment, or broad appeals to corporations, political policymakers, and the scientific establishment—the very interests most responsible for the current ecological mess. In other words, because the crisis has social roots, the solution must involve the transformation of historical relationships on a global scale in order to fashion a sustainable relationship between nature and society.

## THE GROWTH OF THE WORLD ECONOMY

History teaches us that societies have long been at war with the environment, treating nature as little more than a resource to be tapped and as a sink into which to dump their wastes. At times such a one-sided exploitation of nature has led to regional environmental catastrophes that have in turn led to the fall of whole civilizations—the Sumerians are one example that will be discussed in detail later. Nevertheless, until the last few centuries human society existed on so small a scale in relation to the global environment that its effects remained fairly negligible.

This situation began to change with the emergence, in stages, of the capitalist world system, which began in Europe in the late 1400s. The great historical transformation initiated by Columbus' voyage across the Atlantic five hundred years ago marked the origins of what was to become the capitalist world system and the simultaneous creation of a world hierarchy of nation states, defined by the relation of colonizer and colonized, more developed and less developed. The European colonization of the greater part of the globe, beginning with the "New World" and extending to the Asian and African continents, led to the extraction of vast quantities of economic surplus—whether in the form of precious metals, such as gold and silver, or agricultural products, such as sugar, spices, coffee, tea, and many more—and therefore to the social and ecological transformation of the colonized regions. "The gold and silver of the New World," the great French historian Fernand Braudel wrote, "enabled Europe to live above its means, to invest above its savings."[2] The rise of modern colonialism was thus a crucial force behind the commercial revolution experienced by European society from the sixteenth through the eighteenth centuries. Additional sources of food, such as maize, potatoes, and numerous varieties of beans—all of which originated in the Americas—were taken back to Europe, transforming European agriculture and then the agriculture of the entire world. All of this helped prepare the way for the next great stage of capitalism's advance, the Industrial

Revolution, which took place in Europe in the late eighteenth and early nineteenth centuries and led to a rapid increase in the scale and intensity of production and to the development of a set of divisions that are at the core of our understanding of the modern world: the division between economy and nature, between capital and labor, and between center and periphery.

The impact of these developments on the planet's environment becomes clear when we examine changes in four key areas—population, energy, industrialization, and urbanization—over the last four centuries.

## Population

At the time of Columbus' first voyage in 1492 the Americas were the home, according to the most recent estimates, of some 100 million people—compared to a European population of only about 70 million (in 1500). Epidemics and the violence of the conquest led to the rapid decimation of the indigenous Amerindian populations and to the "demographic takeover" of their land by peoples of European origin. "Wherever the European has trod," Charles Darwin observed, "death seems to pursue the aboriginal." Within a century after 1492, the indigenous population had dropped by 90 percent.

This demographic takeover of much of the world by people of European origin was facilitated by a population explosion in early capitalist Europe that was the result of a sharp decline in death rates. This decline was in turn mainly due to the beneficial effects of improved nutrition and sanitation, which reduced famines and epidemics. In England and Wales, the home of the Industrial Revolution, the population was 6 million in 1750. It had increased by half, to 9 million, in 1800. In the next fifty years it doubled, to 18 million. After this, the rate of population *growth* declined, although the population continued to rise—to 33 million in 1900. And the increase took place even though Britain exported 20 million people from the sixteenth century through the early twentieth century (Europe exported 60 million over the same period). As a result of the population explosion at home and expansion abroad, Europeans and their overseas offspring increased their share of world population from about 18 percent in 1650 to more than 30 percent in 1900.[3]

This European population explosion tapered off in the twentieth century. In Europe, as in all regions that have reached fairly advanced stages of economic development, birth rates—responding to higher levels of affluence—eventually fell, producing population growth rates close to replacement level, in the final phase of what is known as the "demographic transition." One of the key characteristics of development, the demographic transition is the gradual changeover from a demographic equilibrium of high death rates and high birth rates, characteristic of preindustrial society, to a demographic equilibrium of low death rates and low birth rates, characteristic of the more developed industrial societies. In the first phase of the transition, associated with early industrialization, there is a decline in the death rate that is not matched by a corresponding decline in the birth rate—and therefore a population explosion. In the second phase, economic development leads to a drop in the birth rate as well, slowing down the rate of population growth. Thus the population of the developed world expanded rapidly in the eighteenth and nineteenth centuries but slowed in the twentieth. By 1985, net population growth in the industrialized countries had dropped to 0.4 percent per year.[4] Affluence thus resulted in lower population growth rates, which then tended to reinforce the affluence.

In contrast, the countries on the periphery of the world economy were unable to follow the same path—to industrialize and then complete the demographic transition. Instead they found their development hindered by the legacy of colonialism and by the need to compete with the advanced capitalist economies, on terms set by the latter. The "development of underdevelopment," as this situation came to be called, was marked by a continual outflow of surplus from the peripheral countries to the core countries, rather than the development of the periphery itself. This resulted in a gap in per capita income between core and periphery that increased sevenfold over the course of a little more than two centuries (we will discuss this further below). The "lag" between the decline in the mortality rate (associated with the move toward industrialization) and the fall in the fertility rate (associated with improved living standards) has thus been lengthened for countries at the bottom of the global economic hierarchy. Caught in a demographic trap between an "industrial" death rate and an "agricultural" birth rate, many underdeveloped countries are faced with an unenviable situation: their populations are growing rapidly while their rates of economic growth per capita are stagnant or declining, each of which then reinforces the other. What has happened, according to the renowned environmentalist Barry Commoner, is that a kind of "demographic parasitism" has developed, in which "the second, population-balancing phase of the demographic transition" in the affluent center of the capitalist world economy "is fed by the suppression of the same phase" in the impoverished periphery. "Colonialism," Commoner writes, "has determined the distribution of both the world's wealth and of its human population, accumulating most of the wealth north of the equator and most of the people below it."[5]

The main consequence of the inability of the periphery to develop economically and complete the demographic transition is that the weight of human numbers has become a growing ecological problem. World population increased from around 500 million in 1600 to more than 750 million in 1750, 1.25 billion in 1850, 2.5 billion in 1950, and 5.5 billion today—an increase of 1,000 percent since 1600. It is expected to exceed 8 billion by the year 2020. And 95 percent of the projected increase will occur in the underdeveloped periphery.

To be sure, the *rate* of population growth is now decreasing because some industrializing countries—such as China, Thailand, and South Korea—have begun to experience the drop in

**An overcrowded urban area in Indonesia contrasts with the modern office towers in the distance. [Sean Sprague/Impact Visuals ]**

the birth rate that defines the second phase of the demographic transition. Nevertheless, the world's population continues to rise and the ability to stabilize it will depend more than anything else on the speed with which the underdeveloped nations are able to reach replacement-level fertility. Long-term United Nations projections suggest that such stabilization will happen when population reaches around 10 billion in 2095, but only if fertility in the periphery (particularly South Asia, Africa, and Latin America) drops to replacement-level by 2035. If replacement-level fertility is not reached until 2065, however, the world population would be 14 billion in 2100. And even this could turn out to be optimistic if there is not further substantive economic development in the periphery, and hence the possibility of completing the second phase of demographic transition in those countries now experiencing a population explosion.[6]

## Energy

The Industrial Revolution was accompanied by a revolution in the conversion of inanimate energy to meet human needs. As the pathbreaking U.S. environmental philosopher and historian Lewis Mumford observed in his *Technics and Civilization* (1934), this was part of a more general shift in the "technological complex" of civilization from the "ecotechnic phase," based on a "water-and-wood complex," to the "paleotechnic phase," based on a "coal-and-iron complex." The growing scarcity of firewood from the sixteenth century on helped spur the demand for coal, and coal production in England rose from 200,000 tons in the mid-sixteenth century to 3 million tons at the end of the seventeenth. This in turn generated the demand for better mine engines, as shafts had to be sunk deeper and deeper below the water table. In the second half of the eighteenth century, James Watt perfected previous innovations and built a steam engine that was widely adopted. It was powered by coal and was at first used primarily in coal mining, but was later used in the textile and metallurgical industries and the railroads. "Coal," British economist W.S. Jevons wrote in 1865, "stands not beside but entirely above all other commodities. It is the material energy of the country, the universal aid, the factor in everything we do. With coal almost any feat is possible or easy; without it we are thrown back into the laborious poverty of early times." World coal production in 1800 amounted to about 15 million tons per year. By 1860 it had risen to around 132 million tons and by 1900 to 701 million (a 46-fold increase over the course of the century).

In the last third of the nineteenth century oil and natural gas began to be exploited, electrical generators were introduced, and giant hydroelectric projects were built. World consumption of petroleum grew rapidly, from 10 million tons in 1890 (shortly after the introduction of the first automobiles) to 95 million tons in 1920, 294 million tons in 1940, and 2.5 billion tons a year in the 1970s. Industrialization was thus associated with the consumption of non-renewable fossil fuels (coal, petroleum and natural gas), supplemented by hydroelectric (and later nuclear) power.

But even as more energy was being produced, more was being sought. World consumption of commercial energy—most of it from nonrenewable sources—rose over 60 times between 1860 and 1985. Moreover, global energy consumption patterns reflected the widening division of the world economy between core and periphery. The average North American today uses 40 times the amount of commercial energy used by the average person in the third world. Energy consumption per person in the advanced capitalist countries is *80 times* that in sub-Saharan Africa.[7]

## Industrialization

Behind these changes in population and energy use lay the fundamental changes in production that are associated with industrialization and the development of capitalism. The Industrial Revolution was dominated in its initial phase (1760-1840) by the rise of the cotton mill in Britain and in its second phase (1840-1875) by the growing use of the steam engine, particularly in the development of the railroad. The next stage (sometimes known as the scientific-technical revolution) began in the late 1800s and involved the development of the steel, chemical, and electricity industries, followed, as the twentieth century approached, by the rise of the automobile. Finally, the second half of the twentieth century saw the growth of petrochemicals, the development of jet aircraft, and the emergence of computers.

Transformations in the division of labor were closely related to each of these changes, progressing from a system of "simple management" under the competitive capitalism of the first Industrial Revolution, to "scientific management" (called Taylorism after its founder Frederick Taylor) under the monopoly capitalism of the scientific-technical revolution, and finally to growing reliance on what is often called "flexible, interchangeable" manufacture under the increasingly globalized production of the present day.

In the context of these changes in production, world manufacturing output (including handicrafts) rose dramatically, increasing more than 80 times between 1750 and 1980 (an estimate that takes into account the effects of the destruction of third world handicraft production during the eighteenth and nineteenth centuries). As recently as 1950, the world manufactured only one-seventh the goods that it produces today and extracted only one-third the minerals. Although the rate of growth has slowed since the 1970s, the world economy continues to expand exponentially: it took only the two decades from 1970 to 1990, when the growth rate of world industrial production was 3.3 percent, for total industrial output to double. With a 3 percent rate of growth, world production will double every 23 years. With a 4 percent rate of growth, it will double every 18 years.[8]

Such exponential growth has been accompanied by growing problems of distribution, both within countries and between center and periphery. In about 1750, per capita income in what are now the developed countries and what are now the underdeveloped countries was roughly equal: about $180 (in 1960 dollars and prices) in the former and between $180 and $190 in the latter. But by about 1930, per capita income in the developed countries had risen to $780 (again in 1960 dollars and prices), while in the underdeveloped countries it had remained at the 1750 level of $190. In 1980, the gap had widened still further, with the developed countries reaching a per capita income of $3,000 and the underdeveloped countries $410. Over the entire period of capitalist development, therefore, the gap between center and periphery grew from 1:1 in 1750 to 4:1 in 1930 to 7:1 in 1980. All of this, in the words of economist Robert Heilbroner, reflects "cumulative tendencies toward strength at the Center to which surplus is siphoned, and weakness in the Periphery from which it is extracted." Seventy-five percent of world production is now absorbed by the 25 percent of the world's population living in the industrial countries.[9]

## Urbanization

Before 1800 the world was overwhelmingly rural, with no more than 2.5 percent of the population living in cities. The first nation to have a level of urbanization above 10 percent was the Netherlands, beginning in the early sixteenth century. The next was Britain. By 1800 London was the largest city in Europe with a population of more than 850,000. By 1985, 43 percent of the world's population was urbanized—although this number reflected wide discrepancies, with urbanization in North America and Europe (excluding the USSR) at 64 and 55 percent respectively, and that of Latin America, East Asia, South Asia, and Africa at 41, 17, 16, and 15 percent respectively.

Large urban agglomerations have environmental effects at three levels: locally, regionally, and globally. Locally, the topography is altered and replaced by a "built environment" of brick, concrete, glass, and metal on, above, and below the ground. Regionally, artificial heat is generated. Cities over 250,000 become "heat islands," hotter than the surrounding countryside (a city of 10 million may have a mean annual minimum temperature as much as 4°F higher than its rural periphery). Globally, these urban areas contribute to atmospheric problems, such as the emission of sulfur and large quantities of carbon dioxide–the latter of which is believed to contribute to global warming.[10]

The greater the division of an economy into urban and rural areas, the greater the spatial concentration of population and industry, leading to concentrated—and therefore more harmful—forms of pollution. Moreover, what economists call the logic of "uneven and combined development"—the unevenness or inequality between regions, together with their combination into a single international division of labor—means that the worst effects of pollution are experienced in third world cities and export enclaves. Air pollution levels are higher in cities in the periphery (or semi-periphery)—such as Bangkok, Bombay, Buenos Aires, Cairo, Calcutta, Manila, Mexico City, Rio de Janeiro, São Paulo, Seoul, and Tehran—than in most of the advanced capitalist world. For instance, the highest allowed ozone level, as set by international standards, is 100 points. On February 6, 1992, Mexico City's ozone level was 342 points and on March 16 it rose to 398 points. Factories were ordered to cut production and schools were closed. In the late 1980s, only 8 of India's 3,119 towns and cities had full sewage treatment facilities and not more than 209 had partial treatment facilities.

Industry-led urban development is accompanied by the spread of "reductionist agriculture" in rural areas, where the capital-intensive and energy-intensive application of hybrid seeds, chemical fertilizers, pesticides, and farm machinery produces cash crops primarily for export. This leads at first to rapid growth in agricultural productivity, but it also creates crops that are increasingly dependent on ever higher, more costly, and less efficient doses of fertilizers and pesticides. Meanwhile, subsistence agriculture is pushed onto marginal or less productive land, leading to deforestation and soil erosion. Among the worst cases is Haiti, where an impoverished population is struggling to survive on deforested mountain slopes while multinational agribusinesses own the most fertile land in the valleys below.[11]

## THE HUMAN TRANSFORMATION OF THE EARTH

The human transformation of the earth has reached a point that is unprecedented in the scale, rate, and kind of environmental change, the cumulative result of the uneven economic development of the last four centuries. This transformation of the earth has reached the point at which there is increasing likelihood of overshooting critical thresholds associated with the finite nature of the ecosphere, with disastrous results for life on earth.

The concept of *critical thresholds*, as Donella Meadows, Dennis Meadows and Jorgen Randers, the authors of the landmark ecological study *Beyond the Limits*, tell us, can be best understood if production is conceived as being made up of a physical "flow or *throughput* from the planetary *sources* of materials and energy through the human economy, to the planetary *sinks* where waste and pollutants end up." This is in accord with the first and second laws of thermodynamics, which tell us that matter and energy cannot be used up but only transformed; that energy dissipates into unusable heat; and that materials cannot be recycled 100 percent.[12] Put simply, the problem is that many of the crucial sources for the streams of energy and materials that feed human society and production are declining, while many of the sinks are overflowing. Renewable sources of energy, including cultivable land, water, forests, and the world's species, are all being exploited at rates that raise questions about sustainability.

At present the world has more food than is needed to feed the world's population. Hunger exists not because of physical limitations but because of the way food is produced and distributed. As the population increases, however, the physical limitations to food production may become increasingly important. For instance, there are definite limits to the amount of potentially cultivable land, estimates of which vary between 2 and 4 billion hectares (depending upon what is considered cultivable), with 1.5 billion hectares already under cultivation and land erosion further reducing the total. Between 1970 and 1990, the world's deserts expanded by about 120 million hectares, more than the amount of land currently cultivated in China. In the years 1970-1990, an estimated 480 billion tons of topsoil was lost, an amount about equal to India's entire cropland. The capacity to increase agricultural yields on available land, meanwhile, may be limited by the reliance on agricultural chemicals—pesticides and fertilizers—in most modern agriculture (a problem that will be discussed later) and by the need for water. It is possible that constraints on agriculture will become critical as the population again doubles part way through the next century. Water is a key consideration in all of this: demand has been rising faster than availability. "Globally water is in great excess, but because of operational limits and pollution," the authors of *Beyond the Limits* argue, "it can in fact support at most one more doubling of demand, which will occur in 20 to 30 years."[13]

Besides being a source of forest-based products, forests form soil, moderate climate, limit floods, and store water against drought. They hold soil on slopes, preventing erosion. They are home to most of the world's species. By taking in and holding carbon, forests stabilize the stock of carbon dioxide in the atmosphere, thereby combatting the greenhouse effect (discussed below). It is therefore significant that half of the world's forest loss over the course of human history occurred between 1950 and 1990. In the fragile ecosystems of the world's tropical forests—where 50 percent of the world's species reside—half of the forest cover is now gone and half of what remains is fragmented and degraded. Even if cutting down tropical forests was limited to a constant 2 percent of the remaining forest per year, which would lead to a smaller and smaller amount cut, it has been estimated that most of the remaining tropical forests would be gone after a century. If the tropical forests are cut down at the same rate at which population is growing in tropical countries, they will be gone in thirty years.

The extinction of species, according to estimates provided by biologist Edward O. Wilson, one of the world's leading experts on biodiversity, is now occurring at the staggering rate of 27,000 species a year—74 every day, 3 every hour. Wilson estimates that up to 20 percent of the world's species could become extinct over the next three decades, a level of extinction not experienced since the disappearance of the dinosaurs 65 million years ago. It has been estimated that there may be 30 million species alive in the world today, only a small proportion of which—1.4 million—

**A lumber train in Oregon. Note that each old-growth log fills a railroad car. [Culver Pictures, Inc.]**

have been identified and catalogued. To lose entire species is not only to reduce the diversity of life on earth, but also to lose the genetic library that might provide new foods, new cancer-fighting drugs, and other products. Human beings now rely on only 20 species for 80 percent of their food supply, yet some 75,000 different edible plant species, some with enormous nutritive value are now known to exist. The winged bean, or *Psophocarpus tetragonolobus* of New Guinea has thus far been ignored by the world's food manufacturers but the whole plant is edible—roots, seeds, flowers, stems, and leaves—and a coffee-like drink can be obtained from its juice. It grows 15 feet tall and has a nutritive value equivalent to the soy bean. Only a few years ago the Pacific Yew tree of the Pacific Northwest was viewed as a "trash tree" to be cut down and burned by loggers. Now it is recognized as the source of the chemical taxol, one of the most important cancer-fighting drugs discovered so far. These are known species, whose value is only just being discovered; millions of other species are threatened with extinction without our ever having learned about them. It is impossible to know therefore what riches—even judged in narrow anthropocentric terms—are being lost.[14]

The depletion of nonrenewable energy sources and materials has also been growing rapidly, although it is impossible to know how far the depletion has gone, particularly with respect to oil, coal and natural gas, since additional deposits of these finite but still abundant resources are continually being located. From an environmental standpoint, at present, the truly serious crises of resource availability have to do with the exhaustion of deep agricultural soil, groundwater, and biological diversity and not with the depletion of fossil fuels or high-grade ores.

In addition to the exhaustion or depletion of planetary sources, we are facing—at the other end of the ecological problem—what can be referred to as "overflowing" global sinks. In ecological terms, a *sink* is the final destination of material or energy flows used by a system of production. For example, the atmosphere is the sink for carbon dioxide from automobile exhaust. A municipal landfill is the sink for paper obtained from trees that grow in a forest. The worst, most intractable pollutants are nuclear wastes, toxic or hazardous wastes resulting from human-synthesized chemicals, and wastes that threaten fundamental global ecological cycles (such as carbon dioxide emissions). All of these forms of waste are growing. According to the authors of *Beyond the Limits*, "Every day 1 million tons of hazardous waste are generated in the world, 90 percent of them in the industrialized world. On an average day in the United States there are five industrial accidents involving hazardous waste." By far the most physical waste is generated by production, not consumption: for every ton at the consumer end of the waste stream there are likely to have been 20 tons in initial resource extraction and 5 tons produced in manufacturing.[15]

No set of pollution problems has so woken us up to the vulnerable state of the planet as the emergence of threats to the atmosphere associated with the depletion of the ozone layer and global warming. The ozone layer in the stratosphere shields the earth from the lethal effects of solar ultraviolet light (UV), which kills cells. An ozone hole was discovered over Antarctica in the 1980s. Later it was discovered that the ozone has been depleted over the northern hemisphere as well. In the spring of 1991, NASA scientists announced that satellite measurements over the northern hemisphere showed that the depletion of stratospheric ozone resulting from the release of chlorofluorocarbons (CFCs) into the atmosphere was occurring at twice the rate previously anticipated. For the first time, these depressed ozone levels extended into the summertime, when radiation damage is most likely to hurt both crops and people. Two-thirds of 300 crops tested are vulnerable to increased UV light, including beans, peas, melons, cabbages, tomatoes, potatoes, sugar beets and soy beans. According to the Environmental Protection Agency, increased UV light is expected to produce 200,000 additional deaths from skin cancer over the next decade in the United States alone. Summer ozone levels fell 3 percent over the northern hemisphere and 5 percent over the southern hemisphere in the 1980s, three times faster than in the 1970s. Recent international agreements have been successful in lowering the level of CFC production. But the Environmental Protection Agency estimates that atmospheric levels of CFCs will increase threefold over the next one hundred years, even if *all* CFC production is stopped by the year 2000, because of the continued leakage of CFCs from old refrigerators, air conditioners, plastic foam furniture, and Styrofoam cups.[16]

The "greenhouse effect" is thought to occur when global emissions of the "greenhouse gases"—carbon dioxide, methane, nitrous oxide, tropospheric ozone gases, and CFCs—produce the global atmospheric equivalent of glass in a greenhouse, allowing light from the sun to enter and heat up the interior but not allowing the resulting heat to pass back through the "glass" (or, in this case, blanket of terrestrial gases). As a result, the earth heats up. Although still uncertain, the postulated greenhouse effect has led scientists to point to "very probable" rises in the global mean temperature of between 1.5° and 5.0°C (2.7° to 9°F) over the next century if "business-as-usual" continues. This would be a degree of climate change unprecedented in the last 10,000 years—a prospect scientists all over the world have presented as potentially catastrophic for the ecology of the entire planet. An increase in average global temperature of 4°C would create an earth hotter than at any time in the last 40 million years. During the last ice age, the average temperature of the earth was only 5°C colder than today, which shows that only a small change in average global temperature can have very large effects. Moreover, the projected warming induced by the greenhouse effect is fifteen to forty times faster than the natural warming at the end of a major ice age, which makes it likely that many species would not have sufficient time to adapt.[17]

All of this can be considered to be part of a much larger problem connected to the ever increasing scale of the human transformation of the earth. In order to understand the ecological turning point now facing humanity, it is essential to recognize the extent to which human activities increasingly rival nature itself with respect to their effect on the fundamental ecological cycles of the planet. Here are some examples of how human activities increasingly rival nature:

---

- **The carbon dioxide content of the atmosphere has grown by 25 percent in the last 200 years, with about half the increase having occurred since 1950. According to Gerard Piel, founder of *Scientific American*, "Furnaces** and internal combustion engines exhaust into the atmosphere 20 percent as much carbon, in the form of carbon dioxide, as is cycled by the two primary life processes: photosynthesis and respiration." It is this fundamental change in the content of carbon dioxide (along with other greenhouse gases) in the atmosphere that is thought to be the cause of global warming.[18]

• The nitrogen cycle, like the carbon cycle, is now being disturbed in significant ways. Nitrogen makes up about 78 percent of the atmosphere, but it cannot be used in this form by plants: it must be made available as ammonia and nitrate through a process of "biological fixation" carried out by bacteria in soil or by aquatic ecosystems. Largely as a result of synthetic fertilizers, humanity now fixes by artificial means about as much nitrogen in the environment as that which is fixed naturally by bacteria associated with legumes. Since nitrate is readily leached from the soil by water, the increased use of nitrogen fertilizer has led to added nitrogen pollution of lakes, rivers, streams, and groundwater.[19]

• Human beings now use (take or transform) an estimated 25 percent of the net photosynthetic product (NPP)—i.e., the plant mass fixed by photosynthesis—over the entire earth (land and sea), and 40 percent on land. As human beings take more of the primary productivity of the earth for themselves, less is left over for other species. To quote the authors of *Beyond the Limits* again, "Somewhere along the path of NPP usurpation, there lie limits. Long before the ultimate limits are reached, the human race becomes economically, scientifically, aesthetically, and morally impoverished."[20]

• The annual human withdrawal of water from natural circulation has increased perhaps as much as 35 times since the late seventeenth century, and has now

**reached approximately 3,600 cubic km. per year, an amount greater than the volume of Lake Huron. In contrast, the annual withdrawal in 1680 has been estimated at less than 100 cubic km.[21]**

- **Since the development of agriculture, net loss of the world's forests as a result of human action is on the order of 8 million square km, an area approximately equal to the continental United States. More than 75 percent of this net loss has been experienced since 1680.[22]**

---

## THE CAUSES OF ENVIRONMENTAL DEGRADATION

What has led to these trends? Environmentalists have put the most obvious causes of ecological degradation into a formula (sometimes known as the PAT formula), which helps us appreciate some of the crucial relationships behind environmental degradation: $I = P \times A \times T$.

In this formula, I is Environmental Impact, P is Population, A is the *material throughput* associated with Affluence (itself defined as capital stock per capita), and T (for Technology) is the environmental impact per unit of energy used to produce the material throughput.[23] By substituting definitions for A and T, the formula can be rendered more precisely as:

Impact = Population × (Capital Stock/Person × Throughput/Capital Stock) × (Energy/Throughput × Environmental Impact/Energy)

The equation shows that environmental degradation is not the result of increased population, *or* increased accumulation, *or* the introduction of less environmentally benign technology. It is a product of all three. Therefore improvements in any of the three variables can have a beneficial environmental impact (and vice versa). Further, as long as both P and A increase, the effect on the environment is bound to be harmful, since T can never be reduced to zero.

In contemporary discourse, environmental problems are most often blamed on Population growth (P), even though levels of Affluence (A) and Technology (T) almost always play a role. Affluence× Technology (AT) is shorthand for the socioeconomic (as opposed to the demographic) causes of environmental degradation, and together they far outweigh the impact of P alone. The throughput (the flow of energy/material input) that supports a wealthy individual, class, or nation is obviously much greater than the throughput that supports a poor individual, class, or nation. If we assume that the use of fossil fuel (both direct and indirect) follows the same pattern as the distribution of income—the higher the income, the greater the use of fossil fuels—then the per capita emission of carbon dioxide for the richest 10 percent of the U.S. population in 1987 was 11 times that of the poorest 20 percent. Similarly, technological choices can be shown to have a vast environmental impact. Barry Commoner has observed that the use of fertilizer nitrogen in the United States increased by 648 percent between 1949 and 1968. Over the same period population increased 34 percent and crop production per capita increased 11 percent, but the use of inorganic fertilizer nitrogen per ton of crop increased a huge 405 percent. Plant roots do not efficiently absorb this added fertilizer and a considerable portion leaches from the soil as nitrate and enters surface waters. The pollution of lakes, rivers, and streams that has resulted from the increased application of nitrogen fertilizer can therefore be traced to the growing reliance on a capital-intensive, energy-intensive technology, and not to population growth or increases in per capita production.[24]

Yet as useful as the PAT formula is for assessing the immediate causes of environmental degradation, it still tells us little about the underlying causes. To ascertain the larger forces influencing P, A, and T, it is necessary to turn to the process of capital accumulation within the center of the world economic system. Thus while environmentalists have sometimes simplistically suggested that the third world can improve its environmental situation by controlling population (P), the East (the former Soviet bloc) by improving on technology (T), and the West by limiting affluence (A), in fact the conditions on which *all* of these factors rest are strongly affected by global dynamics of production, power, and inequality.[25] This means that we must consider the laws of motion of capitalism itself.

## THE SYSTEM

Capitalism today is in its essence what it was at its birth, "a juggernaut driven by the concentrated energy of individuals and small groups single-mindedly pursuing their own interests, checked only by their mutual competition, and controlled in the short run by the impersonal forces of the market and in the longer run, when the market fails, by devastating crises."[26] It is a system of "creative destruction," in which the *creative drive* is the seemingly infinite ability to produce new commodities by combining materials and labor in new ways, and the *destructive drive* is the systematic degradation, transformation, and absorption of all elements of existence outside of the system's own orbit. With the rise of capitalism, environmental historian Donald Worster has noted, humans began to view "everything around them—the land, its natural resources, their own labor—as potential commodities that might fetch a profit in the market. They must demand the right to produce, buy, and sell those commodities without outside regulation or interference."[27] Where the planet is concerned, it is obvious that the rapid growth of capitalism has had overwhelmingly negative results, and that the conservation movements that have arisen since the nineteenth century have failed to restrain the environmental depredations of the system.

Can capitalism be reshaped to the demands of what environmentalists call sustainable development? This is the single most pressing question of our time and one that it will be a major task of this book to answer. If the answer is negative, as I believe it is, then the next question is whether capitalism can be replaced with another kind of society, one that will more fully meet environmental needs. No one can know this for certain. Although there have been many social formations over the course of human history, since the nineteenth century it has been inconceivable that humanity would opt for anything but some form of highly industrialized civilization. And apart from capitalism itself, the only other large-scale model of industrialization was in the former Soviet Union. This was a system that originated under socialist leadership but that was only nominally "socialist" (as this word is understood by most of its advocates, who see these as societies that are controlled by the direct producers) from the 1930s on; that existed in a highly defensive, "cold war" competitive relationship with the advanced capitalist world during its entire existence; and that died out after three quarters of a century. Human experience therefore provides us with little in the way of practical models of ecological and economic sustainability.

Nevertheless, I will argue in what follows that the answers to today's ecological problems do not lie in the direction in which the world is rapidly proceeding—toward the ever greater privatization of nature and the conditions of human existence. Instead they are to be found in the direction of the "socialization" of nature and production, and the creation of a more democratic, egalitarian world order, one that incorporates into its logic an abiding concern for other species and future generations.

But before we can address question of a more ecologically sustainable future, we must take a more detailed look at the reasons why capitalism has failed the environment. Over the last twenty-five years in particular, capitalism has been caught in a double contradiction of economic stagnation and ecological degradation. In order to develop resources for an ecological journey of hope under these dire circumstances, we must explore the interaction between capitalism and the environment over time. "If the environment is polluted and the economy is sick," Commoner writes, "the virus that causes both will be found in the system of production."[28]

# 2

# ECOLOGICAL CONDITIONS BEFORE THE INDUSTRIAL REVOLUTION

Beginning with the development of agriculture 10,000 years ago, all forms of the social organization of production have contributed to the destruction of the environment. Nevertheless, the human relation to nature has not been a uniform one. As production has developed, the relation between nature and society has changed. It is thus possible to distinguish broad "ecohistorical periods"—periods in which "human activities have led to (relatively) uniform changes in nature over vast areas."[1] Such ecohistorical periods can be distinguished by the extent to which human beings have "freed themselves" from subjugation to their environment, on the one hand, and on the other hand by their destructive impact on that same environment.

Viewed in this way, what distinguishes the ecohistorical period of capitalism from the ecohistorical period of precapitalism is not environmental degradation or the threat of ecological collapse—both of which existed before, at least on a regional level—but two traits specific to capitalism. First, capitalism has been so successful over the last few centuries in "conquering" the earth that the field of operation for its destruction has shifted from a regional to a planetary level. And second, the exploitation of nature has become more and more universalized, because nature's elements, along with the social conditions of human existence, have increasingly been brought within the sphere of the economy and subjected to the same measure, that of profitability.[2]

To focus on the environmental impact of distinct ecohistorical periods, differentiating the impact of precapitalist and capitalist eras, is to adopt an historical approach to environmental problems. This represents a break from the mainstream conception, which traces all environmental problems, other than those linked to demographic influences, to the *technology* of production that emerged with the Industrial Revolution. Those who think in this way often disregard the ecological history of humanity prior to the Industrial Revolution, inventing a glorified past characterized by ecological harmony. In the idealized portrait provided by one economic historian:

> Pollution, loss of natural environment, traffic congestion and accidents have clearly resulted from industrialization and modern technology and have no obviously important analogues in preindustrial societies. Moreover, the more work that is done on traditional peasant societies, the clearer does it become that these societies have often achieved an almost miraculous accommodation with nature, balancing present use and preservation for the future with a degree of success which the modern economic machine has rarely approached.[3]

Despite such glorified portraits, it is not just present-day industrial production that presents problems of living sustainably on the earth. Many of our fundamental ecological problems date back to preindustrial times. The modern tendency to blame technology rather than the social systems underlying the technology, moreover, has two results: it discourages any attempt to solve the problem by fundamentally reorganizing society, and it leads to the naive notion that by rejecting industry, and thus simply withdrawing from the modern world, society can return to a mythical state of preindustrial ecological harmony.[4]

None of this is to deny that the environmental impact of preindustrial societies, as measured by Population x Affluence x Technology, was infinitely smaller than it is today. Yet these early societies were especially vulnerable to those regional environmental changes that did take place—often as a result of human interventions designed to extract a larger surplus product—because they raised the specter of ecological collapse whenever the extremely narrow limits of sustainable production were crossed. The history of precapitalist and preindustrial societies is thus full of examples of social collapse brought on by environmental depredations. Historical and archaeological evidence suggests that the Sumerian, Indus valley, Greek, Phoenician, Roman, and Mayan civilizations all collapsed due in part to ecological factors. Finally, the condition of the peasantry, which comprised the bulk of the world's population prior to the Industrial Revolution, was characterized by high infant mortality, low life expectancy, severe undernourishment, and the threat of famines and epidemics—hardly a "miraculous accommodation with nature."[5]

## THE ECOLOGY OF TRIBUTARY SOCIETIES

The first large civilizations were made up of societies that had moved beyond a low level of agricultural development and arrived at a stage characterized by state structures and class hierarchies. Control over the land and its produce was exercised through the extraction of tribute from peasant producers by noneconomic means—hence the designation "tributary societies." Ancient Egypt, feudal Europe, and the Aztec Empire are all examples of tributary societies. Although the timing of its emergence varied across the globe and although social relations differed, the tributary form of production was part of a universal path of development (one that could be found in Europe, Asia, Africa, and the Americas). It constituted the most developed type of economic formation over the course of the more than 5,000 years that stretched from the emergence of Sumer in Mesopotamia, around 4000 B.C., the first literate society in world history, to the rise of capitalism in the late fifteenth century.[6]

Because they were predominantly agricultural, tributary formations were especially vulnerable to ecological collapse stemming from the destruction of the soil. The demise of Sumerian civilization constitutes the earliest recorded example of such ecological overshoot. Sumer emerged in the flood plain of the Tigris and Euphrates rivers and was highly successful for almost 2,000 years. Yet a growing population, the need to feed large numbers of bureaucrats and soldiers—people who no longer grew their own food and who constituted a source of rising demand—and competition between rival cities led to so much irrigation that the conditions of production were slowly undermined. As irrigation increased, the land, which was very flat and drained slowly in part because of the accumulation of silt and mud carried by rivers and canals, became waterlogged and the water table rose. Vast amounts of labor power were then needed to keep the irrigation system clear of the silt and mud. During periods of warfare or internal disorder, the irrigation works fell into disrepair, heightening the tendency of river floodwaters to swamp the region. Salt carried in by the water, combined with the poor drainage and a high rate of evaporation, resulted in the accumulation of a thick layer of salt permeating the topsoil, which destroyed its productivity and eventually led to the collapse of Sumerian civilization in about 2000 B.C.[7]

The decline and fall of the Roman Empire more than 2,000 years later can also be traced to a considerable extent to environmental destruction. Large sections of North Africa, from Tunisia to Morocco, served as a crucial granary for this primarily agricultural empire and were reduced to desert as it became increasingly difficult to produce an adequate amount of food. A growing population, together with the tribute demanded by Rome's ruling strata, led to soil degradation and the extension of production to ever more marginal lands. The cultivation of steep hillsides and vulnerable soils that were easily eroded when deforested and exposed to the elements left the earth desolate. Overgrazing interfered with the natural replacement of pasture, as tree seedlings, grasses, and herbs were destroyed. All of this resulted in erosion so severe that much of the land has still not recovered. The ruins of the great cities of Rome's North African provinces are now surrounded by vast deserts. Meanwhile, silt carried down the rivers in Italy created swamps that became breeding grounds for mosquitoes, leading to severe outbreaks of malaria. Rome thus suffered chronic food shortages and a rising mortality rate after the second century A.D., both of which contributed to a drop in population and a weakening of the empire.

Other ancient tributary formations declined because of the same set of environmental factors. In Mesoamerica, Mayan civilization collapsed around 800 A.D., due in part to extensive tropical deforestation and erosion. An agricultural crisis thus appears to have precipitated the sharp drop in population associated with the Mayan decline. Skeletons dating back to the period immediately before 800 A.D. show higher infant and female mortality and increased levels of disease brought about by declining nutritional standards.[8]

In the feudal society of medieval Europe, which dated from the fall of Rome in the mid-fifth century A.D. until the beginning of mercantilist capitalism in the late fifteenth century, ecological destruction, although in many ways less systematic and devastating than in the Roman Empire, nonetheless continued on a massive scale. Far from a golden age of "natural" peasant cultivation, the medieval period, in the words of English social theorist Raymond Williams, saw "a reduction of most men to working animals, tied by forced tribute, forced labour, or 'bought and sold

**Burning wood for fuel was a major cause of early deforestation, as it is in the third world today. [Sean Sprague/Impact Visuals]**

like beasts'; 'protected' by law and custom only as animals and streams are protected, to yield more labour, more food, more blood.[9] By the end of the eleventh century, only 20 percent of England was still wooded. By 1500, wood consumption in Europe had reached between 60 and 80 million tons, or around 1 ton per person, per year. Depletion of the soil and overgrazing were partly responsible for the famines that recurred "so insistently for centuries on end," in the words of Fernand Braudel, that they became "incorporated into man's biological regime and built into his daily life." Moreover, "famine was never an isolated event. Sooner or later it opened the door to epidemics."[10]

In the late seventeenth century, half the population of France died before the age of twenty and only 10 percent lived to the age of sixty. In the eighteenth century, France experienced *sixteen* famines, each of which stretched across the entire country.[11] As we noted it Chapter 1, it was only the "discovery" of America by the Europeans and the gradual diffusion of New World agricultural products back to Europe during the commercial revolution of the sixteenth through eighteenth centuries that eventually freed Europe from famine: "The introduction and later on the diffusion of the culture of maize and the potato in Europe," Cipolla tells us, "largely contributed, from the eighteenth century onward, to solving the food problem and to reducing the danger of famines when Europe entered a period of accelerated population growth."[12] The "long struggle to survive" that characterized the human condition over most of its early history thus eventually gave way to steady population growth, reflecting fundamental improvements in agricultural productivity and human health.

## CAPITALISM BEFORE THE INDUSTRIAL REVOLUTION

With the emergence of the capitalist world economy, the traditional balance between humanity and nature was irretrievably altered, although at first gradually and only in one small corner of the world. While the new system of production reduced the immediate dependence of society on nature and therefore reduced the threat of ecological collapse, it also opened the way to a vast extension of environmental destruction. The economic, geographic, and scientific changes that characterized the mercantilist stage of capitalism were thus from the outset bound up with the development of a more exploitative relation to the planet.

The "discovery" of the Americas and Vasco da Gama's voyage around the Cape of Good Hope to East Africa and the Indian coast a few years later meant that Europeans now had new worlds to plunder. Reflecting on these conquests , the French philosopher Montaigne wrote at the end of the sixteenth century, "So many goodly cities ransacked and razed; so many nations destroyed and made desolate; so infinite millions of harmelesse people of all sexes, states and ages, massacred, ravaged and put to the sword; and the richest, the fairest and the best part of the world topsiturvied, ruined and defaced for the traffick of Pearles and Pepper." But Montaigne was far more sympathetic to the native populations than most observers. More characteristic of the age was this from philosopher of science and one-time Lord Chancellor of England Sir Francis Bacon: "We have seen what floods of treasure have flowed into Europe by that action. Besides, infinite is the access of territory and empire by the same enterprise."[13]

Nothing so reflected the imperial spirit of the age as the outlook of seventeenth-century science, which saw humanity as engaged in a war for the domination of nature. The conquest of nature, Bacon observed, constitutes "the real business and fortunes of the human race." "By art and the hand of man," he wrote, nature should be "forced out of her natural state and squeezed and molded." Rather than allowing the natural world to continue to dominate humanity, nature must be "bound into service" and made a "slave." For Bacon—as for most powerful figures in his time—the subjugation of nature was to go hand in hand with the subjugation of women.[14]

The first truly global view of the natural world and of the human role in its transformation was that of the great French naturalist Georges-Louis Leclerc, the Compte de Buffon, whose work *Des époques de la nature* appeared in 1779. Buffon described seven epochs in the development of the planet: the first six followed the biblical story of the creation, while the seventh, which he believed characterized the eighteenth century, constituted the age of human dominance. Buffon saw this final epoch as the glorious story of the dominance of human creative powers over nature: "The state in which we see nature today," he observed, "is as much our work as it is hers. We have learned to temper her, to modify her, to fit her to our needs and desires. We have made, cultivated, fertilized the earth; its appearance, as we see it today, is thus quite different than it was in the times prior to the inventions of the arts." "Uncultivated nature is hideous and languishing," Buffon presents the first human as declaring. "It is I alone who can render it agreeable and vivacious."[15]

## The Destruction of Life

The rise of capitalism led to a loss of wildlife on a scale never before seen in human history. The spread of commerce resulted in the death of hundreds of millions of large animals at the hands of traders. In the fifteenth century, sables were common as far west as Finland, but by the late seventeenth century they could only be found in Siberia. By the end of the eighteenth century, nearly every species of fur-bearing animal in Siberia had been decimated and Russian fur traders had to move to the northern Pacific islands, where they killed 250,000 sea otters between 1750 and 1790. With the extermination of fur-bearing animals in western Russia, the fur trade became one of the driving forces behind European expansion into North America. In 1743 the French port of La Rochelle, a center of trade with Canada, imported the skins of 127,000 beavers, 30,000 martens, 1,200 wolves, 12,000 otters and fishers, 110,000 raccoons, and 16,000 bears.

The seal met with the same fate. In the first phase of the world trade, the killing focussed on the southern fur seal. In only seven years (1797 to 1803), over 3 million seals were clubbed to death on the island of Más Afuera in the Juan Fernández Islands, off the coast of Chile. The seal populations of the Falkland Islands and Tierra del Fuego were also destroyed. A total of 6 million southern fur seals were killed in the first two decades of the nineteenth century, with the result that fur seals were virtually extinguished in the Atlantic and Indian oceans.

No example so dramatically shows the broader ecological effects of the extermination of a key species in a region as the beaver. It has been estimated that between 10 and 15 million beavers were killed for their fur in North America in the seventeenth century alone. The New England colonists began to trap beaver soon after they arrived, and by the late 1600s the beaver was already scarce in most of New England. By the late 1700s, the beaver population had been extinguished east of the Allegheny Mountains. Exported to England as pelts, beaver fur was often re-exported back to the colonies (particularly those with large commercial elites) in the form of hats. The loss of the beavers, with their dams and ponds, led to a drastic change in the ecology of the region and this in turn affected other species. Fewer and fewer black ducks, ring-necked ducks, hooded mergansers, and goldeneyes returned to breed in beaver ponds. With the beavers no longer maintaining their dams, muskrat and otter were flooded or frozen out by fluctuating pond levels. Mink and raccoon, which ate the frogs, snakes, and suckers in beaver flowages, were cut off from this part of their food supply when ponds shrank into marshes and were eventually transformed into meadows. Tree stumps left by the beaver had once sprouted tender stalks and leaves that had attracted rabbits and snowshoe hare. Trees felled by the beaver had once provided the brush that had protected the rabbit and drumming logs for the springtime mating of the ruffed grouse. As a result, red foxes, which depended on these animals for their food supply, found fewer to stalk. The commercial trade in beaver hats thus contributed to the destruction of an entire ecosystem—a process that continued as the beaver trade moved west across the United States and Canada.[16]

## Sugar and Slavery

The war on other species reflected the dominance of commercial ends. The ecological effects of the mercantilist age of capitalism, however, were to be found not simply in the destruction of animal species for profit, but in the creation of a world system of cash-crop production based on the transformation of nature and the subjugation of human labor.

The first cash crop to transform the environment of the periphery was sugarcane. The main elements of the sugar plantation system were already in place in the Atlantic islands of Madeira and the Canaries in the mid-fifteenth century, where sugar production based on slave labor first took hold. The slaves were drawn from among the native inhabitants of the Canaries—the Guanches—who numbered around 80,000 at the time of the conquest of the islands but had been completely killed off by 1600, and also from North Africa.[17]

Christopher Columbus, who was acquainted with the Atlantic islands, seems to have intended to reproduce this same system of slave-based plantation agriculture on Hispaniola, and when he set out on his second voyage in 1493 he took along several stalks of sugarcane. On Hispaniola the hard labor on plantations and in mines was carried out by Amerindian slaves. It was not until 1518, after the Indian population had been decimated, that the mass importation of slaves from Africa began. But as the demand for sugar grew in the sixteenth and seventeenth centuries, so did the demand for slaves. Sugar production quickly spread throughout the Spanish colonies and into the Portuguese colony of Brazil, which quickly became the main sugar-producing region in the world, and from there to the British and French colonies in the Caribbean, which displaced Brazil.[18]

Between 1451 and 1600, some 275,000 African slaves were sent to America and Europe. In the seventeenth century, this number rose to an estimated 1,341,000, largely in response to the demand of the sugar plantations in the Caribbean. It was the eighteenth century, however, that was to be the golden age of slaving, with the forcible exportation of more than 6 million people from Africa to the Americas between 1701 and 1810.[19]

A double genocide lay behind this vast demographic change. "Twelve years after Christopher Columbus landed in this part of the world," novelist Jamaica Kincaid, herself a West Indian, has written, "over a million of the people he found living here were dead. In addition, so many Africans were thrown overboard on voyages from Africa to this part of the world that it would not be an overstatement to say that the Atlantic Ocean is the Auschwitz of Africa."[20]

The land too became a "slave" to the new system of export crop production. Planters cleared it of trees, making it more prone to drought and erosion. In 1690 trees still covered more than two-thirds of Antigua; by 1751 every acre suitable for cultivation had been stripped of forest cover. Intensive cultivation of sugarcane mined the soil, robbing it of its nutrients.[21] What Uruguayan writer Eduardo Galeano wrote of the Northeast of Brazil was true of most of the Caribbean islands as well:

> Sugar ... destroyed the [Brazilian] Northeast.... This region of tropical forests was turned into a region of savannas. Naturally fitted to produce food, it became a place of hunger. Where everything had bloomed exuberantly, the destructive and all-dominating latifundio left sterile rock, washed-out soil, eroded lands. At first there had been orange and mango plantations, but these were left to their fate, or reduced to small orchards surrounding the sugarmill-owner's house, reserved exclusively for the family of the white planter. Fire was used to clear land for canefields, devastating the fauna along with the flora: deer, wild boar, tapir, rabbit, pacas, and armadillo disappeared. All was sacrificed on the altar of sugarcane monoculture.[22]

The creation of a sugar monoculture left these colonies dependent on Europe, North America, and the South American interior for their food. "To feed a colony in America," Abbé Raynal ironically observed in 1775, "it is necessary to cultivate a province in Europe." At the end of the sixteenth century, reports Galeano, "Brazil had no less than 120 sugarmills worth some £2 million, but their masters, owners of the best lands, grew no food. They imported it, just as they imported an array of luxury articles which came from overseas with the slaves and bags of salt."[23]

The export of sugar grew rapidly. After 1660, England's sugar imports exceeded its combined imports of all other colonial produce; by 1800 the English population consumed almost 15 times as much sugar as it had in 1700. Yet sugar, despite its importance, was only one pillar in the triangular trade that linked

Europe, Africa, and the Americas. The first leg in the triangle was from a European port to Africa, in ships that carried a cargo of salt, textiles, firearms, hardware, beads, and rum. These products were bartered for slaves, who were packed into the vessels like "rows of books on shelves" (with each individual having a space as small as five and one-half feet long and sixteen inches wide) for the voyage to the Americas. There, the slaves were sold to plantation owners, and in the last leg of the triangle, sugar, silver, molasses, tobacco, and cotton—all of which had been produced with the help of slave labor—were purchased and shipped back for sale in Europe. In Britain such important seaports as Liverpool, Bristol, and Glasgow owed their rapid growth in the eighteenth century primarily to the triangular trade.[24]

### Soil and Civilization

In the seventeenth and eighteenth centuries planters in Virginia and Maryland grew tobacco with slave labor. Virginia's settlers had cleared some half million acres of forest by the end of the seventeenth century. According to historian Ralph Davis, this destruction was followed, within three or four years, by "the draining away of much of the best soil; and tobacco-cropping very rapidly exhausted the land's fertility. Every planter, therefore, tried to build up a large landholding with plenty of uncleared land in reserve, using the ample spare labour of the slack season each year to clear a piece of virgin land to which he moved some of his tobacco cultivation."[25]

As a result, land prices soared and the area of cultivation spread rapidly westward, across Virginia and Maryland and into what were then outlying territories. The pressure of the colonists, and of land speculators (like George Washington and Benjamin Franklin), led to a major Indian rebellion (miscalled "Pontiac's Conspiracy" by the British) in 1763—following the French defeat in the French and Indian war (1756-1763). At the instigation of the Delaware and Seneca, more than a dozen Indian nations resisted the colonial invasion, attacking British outposts throughout "Ohio country," which stretched north-south from Lake Erie to the Ohio River and east-west from the Allegheny Mountains to western Ohio. In order to bring the fighting to an end, the British King issued the Proclamation of 1763, which said that no more European settlements could be established west of the Appalachians. This did little to halt land speculation, however. Thus plantation owner George Washington secretly employed a surveyor to locate more land to the west. He wrote to a friend that "I can never look upon that proclamation in any other light (but this I say between ourselves) than as a temporary expedient to quiet the minds of the Indians...."[26]

The expansion westward led to further conflict with the Indians. The most important and extensive political unit among the Indians north of the Aztec Empire was the Iroquois Confederacy, which was based in upstate New York and western Pennsylvania but extended into the Ohio country. The Confederacy included six Indian nations that were bound together by a common Iroquois language—the Mohawk, Onondaga, Seneca, Oneida, Cayuga, and (after they were driven out of North Carolina as a result of attacks by colonial armies in the early eighteenth century) the Tuscaroras. The Iroquois held their land in common, hunted in common, and had houses (shared by several families) that were considered common property. In the 1650s a French Jesuit priest observed: "No poorhouses are needed among them, because they are neither mendicants nor paupers.... Their kindness, humanity and courtesy not only makes them liberal with what they have, but causes them to possess hardly anything except in common."[27]

The territory dominated by the Iroquois (particularly the Ohio country) was the core of a region coveted by both the French and the British, and the Iroquois Confederacy became adept at playing the two colonial powers off against each other—until the French were forced out of North America by the British in 1763. With the disappearance of the French from the North American scene, the fate of the Indians, who could no longer play off one colonial power against another, was sealed. In the Revolutionary War of 1776, the Iroquois were divided, with some nations supporting the British and some the colonists. In retaliation for Iroquois attacks, General Washington, who later referred to the Indians as "beasts of prey," ordered a campaign of total destruction.[28] On May 31, 1779, he commanded General Sullivan:

> The expedition that you are appointed to command is to be directed against the hostile tribes of the six nations of Indians... The immediate objects are the total destruction and devastation of their settlements and the capture of as many prisoners of every age and sex as possible. It will be essential to ruin their crops now on the ground, and prevent their planting more.... [P]arties should be detached to lay waste all the settlements around, with instructions to do it in the most effectual manner, that the country may not be merely overrun but destroyed.[29]

Crops such as corn, beans, potatoes, pumpkins, squash, cucumbers, and melons grew in an abundance that astonished the invading soldiers. Some of the corn stalks were sixteen feet high and the ears as much as twenty-two inches long. There were also apple, peach, and cherry orchards. The orchard in one town contained 1,500 fruit trees. None of this was left intact. Forty towns and scattered settlements containing large houses were burned. In his report Sullivan declared, "We have not left a single settlement or field of corn in the country of the Five Nations [sic], or is there even the appearance of an Indian on this side of the Niagara." Before the Revolutionary War most of the population of the Six Nations lived in thirty thriving towns scattered from the Mohawk River to Lake Erie in the Ohio country. By the spring of 1780, only two of these towns were undamaged by the war; the others were empty ruins or lay in ashes. The population was cold and hungry, suffering from scurvy and dysentery. Washington's role in this destruction was not forgotten by the Iroquois. In 1792 the Seneca Chief Cornplanter addressed President Washington as follows: "When your army entered the country of the Six Nations, we called you the Town Destroyer; and to this day, when that name is heard, our women look behind them and turn pale, and our children cling close to the necks of their mothers."[30]

The Iroquois fought back, but when the British were defeated in 1783 they found themselves treated as conquered subjects by the new Americans. According to historian Anthony Wallace, "Land speculators ... saw the possibility of vast profits to be made from the sale, to thousands of individual settlers and entrepreneurs, of virgin timber and agricultural land, of waterways, mill sites, trading locations, harbors, towns sites, and so on." Within thirteen years the Iroquois had been deprived of nearly all of their land and forced onto a few small reserves.[31]

In New England, declining soil fertility had created a hunger for Iroquois land. In Concord, New Hampshire, grain yields had fallen from 13.2 bushels per acre in 1749 to 12.2 bushels in 1771, while meadow hay yields fell from 0.82 to 0.71 tons per acre. In 1749 it took 1.4 acres of pasture to support a cow; by 1791 it took 4.1 acres. Colonial agriculture, with its reliance on a single crop, was vastly inferior to Indian agriculture, in which the practice was to plant corn, beans, and squash together, the corn stalks acting as bean poles and shading vegetables below. Monoculture not only meant lower yields per acre but also depleted the soil. As the Indians were pushed off their lands in upstate New York, Massachusetts farm families flooded westward. The expansion of the land base available to European settlers, as Carolyn Merchant has argued, was one of the crucial factors in the "ecological revolution" that took place in New England and New York at the end of the eighteenth century—the shift from coastal mercantilism and subsistence farming to capitalist agriculture and industry. The former had relied primarily on "extensive" methods of increasing production; the latter substituted "intensive" techniques of soil and crop management, while at the same time increasing the overall demand for land.[32] The forcible removal of the original inhabitants—in this case the Iroquois—was therefore a form of "primitive accumulation" that set the stage for the development of a wider, more intensive capitalist economy.

# 3

# THE ENVIRONMENT AT THE TIME OF THE INDUSTRIAL REVOLUTION

The mercantilist period saw the development of a commercial, agrarian, and mining capitalism in England, which by the eighteenth century had replaced the Netherlands as the most advanced capitalist economy. The proceeds from the trade in spices, sugar, tea, coffee, tobacco, gold, furs, and slaves fed profits into a post-feudal English social order that was manifested in the rural areas by what Raymond Williams called "the country-house system of the sixteenth to eighteenth centuries."

This system, in which landlords in great houses owned vast estates run by tenant farmers and worked by agricultural wage-laborers, had been made possible by the agricultural revolution that had taken place under mercantilism. England was notable among European countries in that the traditional peasantry disappeared early, primarily as a result of the enclosure movement. The percentage of English agricultural land that was *enclosed* by stone walls and hedges—so as to be more systematically monopolized by rural landlords—rose from 47 percent in 1600 to 71 percent by 1700. A further 6 million acres were enclosed in the eighteenth century. By the end of the seventeenth century, 40 percent of the English population had moved out of agricultural employment, mostly into industrial pursuits. By the end of the eighteenth century, nearly half the cultivated land in England was owned by 5,000 families, while nearly 25 percent was owned by a mere 400 families.

This system, despite the inequality built into it, was more productive than anything that had preceded it. The agricultural revolution, by introducing new crops, improving methods of cultivation, and bringing additional land under cultivation, increased domestic agricultural output in England between 1550 and 1750 sufficiently to feed a population that doubled over the same period, while at the same time reducing the share of agriculture in total employment. Large numbers of workers were "liberated" from the land and sought employment in the industry of the towns.[1]

These conditions thus set the stage for the Industrial Revolution, which began in the late eighteenth century. The Industrial Revolution added a new intensity to capitalism's relation to its environment. Although the commercial and agricultural revolutions of the mercantilist period had begun to alter the human relationship to the earth on a global scale, mercantilism was mainly an extensive phase of development, working its changes more by a process of ecological takeover than ecological transformation. It was the rise of machine capitalism that made possible the real subjection of the original sources of wealth—the soil and the worker—to capital. Driven by its inner logic to commodify such essential elements of industry as land and labor, yet unable to do so without undermining the natural and human bases of existence, capitalism found itself more and more at war with its environment.

The fact that the Industrial Revolution had adverse ecological effects was understood from the very beginning. Surveying the iron works around Coalbrookdale in 1830, James Nasmyth, the inventor of the steam hammer, wrote: "The grass had been parched and killed by the vapours of sulphureous acid thrown out by the chimneys; and every herbaceous object was of a ghastly grey—the emblem of vegetable death in its saddest aspect." It is therefore surprising that it is sometimes suggested today that the harsh conditions of the Industrial Revolution did not lead to the immediate growth of environmental criticism. Thus political ecologist Hans-Magnus Enzensberger has argued that the contamination of the environment associated with "English factories and pits" should have provided "food for ecological reflection. But there were no such observers. It occurred to no one to draw pessimistic conclusions about the future of industrialization from these facts." In this he was mistaken. Poets, novelists, journalists, physicians, Romantic social analysts, and defenders of the working class gave eloquent testimony to the horrors of the new industrial system. For the great English poet William Blake, the question was posed as follows:

> And was Jerusalem builded here
> Among these dark Satanic Mills?[2]

The political economists of the period introduced what was to become the classic debate on the relations between overpopulation, poverty, and environmental degradation. Two broad positions were put forward, one associated with the name of Thomas Malthus and the other with that of Karl Marx. Even today, as the renowned environmental economist Herman Daly has written, "The Marxian and Malthusian traditions represent the major competing explanations of poverty in Western thought," without which modern environmental problems cannot be addressed.[3]

England's Midlands were called the "Black Country" because so much of the vegetation had been destroyed. [Culver Pictures, Inc.]

## DARK SATANIC MILLS

The Industrial Revolution can be defined as a sudden take-off in growth as the result of a series of economic, social, and ecological transformations. Its principal elements were the growth of the factory system, the expansion of wage labor, the increased reliance on machine production, and the rise of the modern industrial city—symbolized above all by the English city of Manchester.[4]

For those who witnessed the emergence of this new stage of production, it was the contrast between the enormous riches produced by this system and the deterioration of environmental conditions that was most shocking. "From this foul drain," the noted French social analyst Alexis de Tocqueville wrote of Manchester in 1835, "the greatest stream of human industry flows out to fertilize the whole world. From this filthy sewer pure gold flows. Here humanity attains its most complete development and its most brutish, here civilization works its miracles and civilized man is turned almost into a savage."[5]

At the center of the Industrial Revolution was the cotton industry. The British cotton industry was originally an outgrowth of overseas trade, which provided both its raw material and the bulk of its final product in the form of calicoes imported from India by the East India Company. Soon, however, the British wool industry managed to secure import prohibitions against Indian calicoes, giving British cotton manufacturers, who were originally uncompetitive, a chance to develop. This facilitated the rapid growth of the internal market for cotton goods in Britain, although the main growth market for these goods would soon be found in the British colonies and overseas dependencies.

Colonialism thus launched the British cotton textile industry and remained a key to British expansion throughout the Industrial Revolution. The cotton-goods industry's fastest growth in the eighteenth century was on the outskirts of the major colonial-trade ports of Bristol, Glasgow, and particularly Liverpool, the great hub of the slave trade. And this was no coincidence: during the entire first phase of the Industrial Revolution (up until 1840) "slavery and cotton marched together," as British historian Eric Hobsbawm put it. Prior to the industrial take-off, the greater part of Lancashire's cotton exports went to markets in Africa, where they were exchanged for slaves, and to the Americas, where West Indian sugar planters bought cotton goods in large quantities. By the 1790s, the insatiable needs of the British cotton mills had created a rocketing demand for the raw cotton produced on the slave plantations of the U.S. South.

Between 1750 and 1769 British cotton goods exports rose ten

times over. The great bulk of these exports were to colonial and semi-colonial regions. In 1840, Europe took 200 million yards of English cotton exports, while the Americas outside of the United States, Asia, and Africa took 529 million yards. Beginning with the Napoleonic wars, Latin America became an economic dependency of Britain and by 1840, despite its poverty, was absorbing one and a half times as many British textiles as Europe. India, which had been a traditional manufacturer and exporter of cotton textiles, was deindustrialized under British colonial rule. In 1820 the Indian subcontinent took 11 million yards of British cotton textiles; by 1840 this had grown to 145 million yards.[6]

"The division of labour," Adam Smith observed in *The Wealth of Nations,* is "limited by ... the extent of the market."[7] The fact that the market for cotton was from the start a global one meant that the prospects for economic expansion were enormous, making rapid changes in specialization and the division of tasks within the workplace possible. This allowed for the emergence of the modern factory, dominated by machine production, which up until the mid-nineteenth century was largely identified with the cotton goods industry.

Cotton manufacture led to urbanization, and what happened in the British city of Manchester is only an extreme case of what was happening in other cities. The population of the city rose more than tenfold, from 17,000 to 180,000, between 1760 and 1830, presenting a view of "hundreds of five- and six-storied factories, each with a towering chimney by its side, which exhales black coal vapour."[8]

*In Hard Times* (1854) novelist Charles Dickens describes an "imaginary" Coketown as follows:

> It was a town of red brick, or of brick that would have been red if the smoke and ashes had allowed it.... It was a town of machinery and tall chimneys, out of which interminable serpents of smoke trailed themselves for ever and ever, and never got uncoiled. It had a black canal in it, and a river that ran purple with ill-smelling dye, and vast piles of buildings full of windows where there was a

> rattling and a trembling all day long, and where the piston of the steam-engine worked monotonously up and down like the head of an elephant in a state of melancholy madness.

Dickens depicts Stephen Blackpool, the working-class protagonist of *Hard Times*, as living "in the innermost fortifications of that ugly citadel, where Nature was as strongly bricked out as killing air and gases were bricked in."[9]

The conditions of work were appalling: "The spinners in a mill near Manchester had to work fourteen hours a day in a temperature of eighty to eighty-four degrees without being allowed to send for water to drink."[10] Such extreme exploitation, arising from the systematic attempt to increase the profits of capital, led to a degree of ill-health and physical deformation that were among the most visible results of factory labor.

Many women and children were employed in the factories during the Industrial Revolution because they could be paid far lower wages than adult males. Indeed, in his elaboration of the principles of the division of labor, the early nineteenth-century management theorist Charles Babbage explained that behind the division of work into different tasks lay the desire to simplify individual work processes so that labor of cheaper varieties (Babbage used the example of women and children) could be substituted for labor of the more expensive varieties. The substitution of child for adult labor wherever possible was particularly economical since children generally received only one-third, and sometimes as little as one-sixth, of the adult wage. As the famous historian of the Industrial Revolution Paul Mantoux reports, "The first Lancashire factories were full of children. Sir Robert Peel had over a thousand in his workshops at once.... Lots of fifty, eighty or a hundred children were supplied [by the parishes] and sent like cattle to the factory, where they remained imprisoned for many years." Yet there were many defenders of the factory system. For instance, when physicians called before a factory investigation committee testified that exposure to sunlight was essential to the physical development of children, Andrew Ure, a leading exponent of the principles of manufacturing, replied in indignation that the gas lighting of the factory was an adequate substitute for the sun.[11]

If the factory environment during the Industrial Revolution was grim, the larger urban environment was even more so. Factory workers across England lived in squalor and were plagued by hunger and disease. In the first-hand description provided in his *Condition of the Working Class in England* (1845), Frederick Engels (who was to become Karl Marx's lifelong intellectual collaborator) walked the reader through whole areas of Manchester, street by street, describing what was to be seen and arguing that the environment of the streets occupied by the working class was so different from that of the bourgeoisie as to constitute two different worlds. The homes of the "upper bourgeoisie" of Manchester were to be found "in remoter villas with gardens in Chorlton and Ardwick, or on the breezy heights of Cheetham Hill, Broughton, and Pendleton, in free, wholesome country air, in fine comfortable homes passed once every half hour or quarter hour by omnibuses going into the city. And the finest part of the arrangement," Engels observed, "is this, that the members of this money aristocracy can take the shortest road from the middle of all the labouring districts to their places of business, without ever seeing that they are in the midst of the grimy misery that lurks to the right and the left."[12]

In surveying the conditions of the working class in London, Manchester, and elsewhere, Engels was particularly concerned with environmental toxins. Relying on the reports of physicians and on his own personal observations, he provided a detailed analysis of public health conditions. Using demographic data compiled by public health officials, he pioneered in arguing that mortality rates were inversely related to social class, which could be seen most dramatically by examining specific sections of each city. The poorly ventilated houses of the workers, he argued, did not allow for adequate ventilation of toxic substances, and carbon gases from combustion and human breathing remained trapped inside. Since there was no system for the disposal of human and animal waste, these accumulated and decomposed in apartments, courtyards, and streets, producing severe air and water pollution. The high mortality from infectious diseases, such as tuberculosis (an airborne disease) and typhus (carried by lice), was the result, he argued, of overcrowding, bad sanitation, and insufficient ventilation.

---

**In 1845 a young (24-year-old) Frederick Engels gave the following first-hand account of living conditions in industrial Manchester: "In a rather deep hole, in a curve of the Medlock and surrounded on all four sides by tall factories and high embankments, covered with buildings, stand two groups of about two hundred cottages, built chiefly back to back, in which live about four thousand human beings, most of them Irish. The cottages are old, dirty, and of the smallest sort, the streets uneven, fallen into ruts and in part without drains or pavement; masses of refuse, offal and sickening filth lie among standing pools in all directions; the atmosphere is poisoned by the effluvia from these, and laden and darkened by the smoke of a dozen tall factory chimneys.... The race that lives in these ruinous cottages, behind broken windows, mended with oilskin, sprung doors, and rotten door-posts, or in dark, wet cellars, in measureless filth and stench, in this atmosphere penned in as if with a purpose, this race must really have reached the lowest stage of humanity. This is the impression and the line of thought which the exterior of this district forces upon the beholder. But what must one think when he hears that in each of these pens, containing at most two rooms, a garret and perhaps a cellar, on the average twenty human beings live; that in the whole region, for each**

**one and twenty persons, one usually inaccessible privy is provided.... Dr. Kay asserts that not only the cellars but the first floors of all the houses in this district are damp; that a number of cellars once filled up with earth have now been emptied and are occupied once more by Irish people; that in one cellar the water constantly wells up through a hole stopped with clay, the cellar lying below the river level, so that its occupant a hand-loom weaver, had to bale out the water from his dwelling every morning and pour it into the street!"[13]**

---

Engels also described the skeletal deformities caused by rickets as a nutrition-related problem, even though the specific dietary deficiency associated with this, the lack of Vitamin D, was not yet known. He provided accounts of occupational illnesses, including detailed descriptions of orthopedic disorders, eye disorders, lead poisoning, and black lung disease.[14]

It was "plain," Lewis Mumford wrote in his magnificent study *The Culture of Cities* (1938), "that never before in recorded history had such vast masses of people lived in such a savagely deteriorated environment." Epidemics of cholera and typhoid took an appalling toll in the years after 1830.[15]

## MALTHUS AND POPULATION

For the well-to-do of the nineteenth century, these conditions were seen as part of the "population problem," which they decried as an ill but at the same time used to argue that it made the harsh environment of capitalist industrialism necessary—since, as one of them put it, where else "could the millions by which the population of England has increased find work?"[16] Prior to 1700 the increase in population over the course of *each* one hundred years was about 1 million. Between 1700 and 1800, the increase was 3 million. The industrial takeoff gave added momentum to this, making England a center of both industrialization and rapid population expansion.[17]

In 1798 the classical economist Thomas Robert Malthus published his *Essay on the Principle of Population.* His central theme was that the vast majority of the population faced extreme poverty, and that efforts to remedy this situation would do more harm than good. Anticipating the views of many of today's ecologists, Malthus argued that all animals, including human beings, had the capacity to increase geometrically (1, 2, 4, 8, 16, and so forth). The human population, if unchecked, would thus be expected to increase "in a geometrical progression of such a nature as to double itself every twenty-five years."[18] But since the world's population had rarely if ever increased at this "natural rate," the greater part of Malthus's analysis was concerned with analyzing the forces that had held it in check.

The most important of these was the limitation on the food supply within any given territory. Although the food supply would, Malthus contended, have a tendency to increase as a result of the application of additional labor, the extension of the land under cultivation, and improvements in agricultural technique, the increases in supply from one generation to another would nevertheless tend to diminish, largely because all of the best land would eventually be brought under cultivation. Food production would thus increase only arithmetically (1, 2, 3, 4, 5, and so on) at best. Without any other checks, therefore, starvation would limit population growth to the rate at which the food supply could be increased. But there were additional checks: preventative checks, those that reduced the birth rate and included sterility, sexual abstinence, and birth control; and positive checks, those that increased the death rate and included misery, plagues, and war (along with famine).

Both preventative and positive checks maintained an equilibrium between population growth and the available means of subsistence. Ultimately, Malthus argued, they boiled down to "moral restraint, vice and misery." He defined moral restraint as "the restraint from marriage which is not followed by irregular gratifications." Perhaps the most important distinction between the rich and the poor, he believed, was the greater moral restraint of the former: "Carelessness and want of frugality," vice and misery, he argued, were common among the impoverished elements of society. For these reasons Malthus opposed all measures that would alleviate the harsh impact of the market on the poor.[19]

In expert testimony before a Parliamentary Committee on Emigration in 1827, Malthus advocated reforms in the Poor Laws that would create an even harsher environment for those seeking parish relief and urged the committee to refuse relief to all children who were born two years after his reforms had been instituted. He also commended landlords who pulled down cottages the moment they became vacant, and argued against the construction of new cottages, since he believed that a shortage of housing would discourage early marriage.[20]

Although Malthus' views on overpopulation were to influence many later ecologists, Malthus himself was not particularly interested in, or even aware of, what would nowadays be called the larger ecological implications of his analysis (i.e., the carrying capacity of planetary ecosystems). Rather, he was engaged in a dispute with radicals such as William Godwin (1756-1836), who had been sympathetic to the ideals of the French Revolution and who advocated humanitarian measures to ameliorate the worst hardships of the poor. For Malthus, the population doctrine showed conclusively that "the future improvement of society" could not occur through any process that reduced inequality.[21]

The weakest part of the Malthusian doctrine was the purported "arithmetical ratio" governing food supply. Malthus provided no evidence for its existence, merely asserting that "by great exertion, the whole produce of the Island might be increased every twenty-five years, by a quantity of subsistence equal to what it at present produces. The most enthusiastic speculator cannot suppose a greater increase than this."[22] In his later work he placed greater emphasis on the role played by the law of diminishing returns in agriculture, the result of the gradual extension of cultivation to physically inferior land. This led him to even more pessimistic conclusions. Once all the land was occupied, he wrote, "the rate of increase of food would ... have a greater resemblance to a decreasing geometrical ratio than an increasing one. The yearly increment of food would, at any rate, have a constant tendency to diminish."[23]

## MARX AND RELATIVE SURPLUS POPULATION

Malthus based his analysis of overpopulation, poverty, and the decline of the environment on presumed "natural laws" of food production and human fertility. Karl Marx attributed such problems primarily to social causes. Marx contended that "every particular historical mode of production has its own special laws of population, which are historically valid within that particular sphere. An abstract law of population exists only for plants and animals, and even then only in the absence of historical intervention by man." "Overpopulation," he wrote, "is ... a historically determined relation in no way determined by the absolute limit of the productivity of the necessaries of life, but by limits posited rather by *specific conditions of production* [along with] the *conditions of reproduction of human beings.*"[24]

Under industrial capitalism the most important law governing population, Marx argued, is that of relative surplus population or the industrial reserve army of the unemployed. Wages (or historically defined subsistence levels) are not determined by the relation between population and food but by the relation between population and *employment.* The existence of a mass of unemployed is the lever capitalism uses in its efforts to reduce labor power (an individual's capacity to work) to the status of a commodity—a good to be bought and sold like any other. But this law of relative surplus population does not hold for all stages of capitalism: it only comes into being with the rise to dominance of machine production. Before this could happen, capitalism had to alter labor's relationship to the land by making the land itself into a good to be bought and sold. This—the expropriation of the land, which made possible the expropriation of the agricultural laborer—was the real secret of the English agricultural revolution that preceded the Industrial Revolution.

The presence of an industrial reserve army leads to increased competition among workers for the limited amount of employment, and this in turn holds down wages. Wages are thus determined not by an iron law revolving around the price of grain (as earlier economic thinkers, such as Adam Smith and Malthus, had argued), but by historically determined standards of subsistence and by "the respective power of the combatants" in the class struggle. The industrial war, Marx argued, differs from other forms of modern warfare in that "the battles in it are won less by recruiting than by discharging the army of workers." He referred to the law of relative surplus population as "the absolute general law of capitalist accumulation," since it is the "lever" for the concentration of relative wealth at one pole and relative poverty at another.[25]

Marx believed that not only labor but nature too was increasingly being subjected to capital as a result of the new conditions ushered in by the Industrial Revolution. By separating town and country, and by applying industrial techniques to the latter as well as the former, capitalism disrupted the ecological basis of human existence:

> All progress in capitalist agriculture is a progress in the art, not only of robbing the worker, but of robbing the soil; all progress in increasing the fertility of the soil for a given time is progress towards ruining the long-lasting sources of that fertility. The more a country proceeds from large-scale industry as the background of its development, as in the case of the United States, the more rapid is this process of destruction. Capitalist production, therefore, only develops the techniques and the degree of combination of the social process of production by simultaneously undermining the original sources of all wealth—the soil and the worker.[26]

These were not offhand comments but reflected careful study of the work of the German agrarian chemist Justus von Liebig, often known as the founder of soil chemistry. The intensification of agriculture during the first phase of the Industrial Revolution had led to a progressive loss of fertility and a drop in agricultural yields in some regions. The desperation of the farmers was so great that some retrieved animal and human bones from the battlefields of the Napoleonic wars to spread over their land. Liebig had been a major advocate of the importation of Latin American guano for fertilizer. In 1842 the first artificial fertilizer was introduced by the English agricultural chemist John Bennett Lawes, who devised a means of making phosphate soluble and established the first fertilizer factory. But it was not until just before World War I that the German chemist Fritz Haber developed a way to make an artificial nitrogen fertilizer.[27]

Until the early 1860s, Marx, who rejected the classical liberal theory of diminishing returns in agriculture, thought that the progress of capitalist agriculture might be so rapid that it would outpace industry. But by the time he wrote *Capital,* his studies of the work of Liebig and other agronomists had convinced him otherwise. "Large landed property," he explained,

> reduces the agricultural population to an ever decreasing minimum and confronts it with an ever growing industrial population crammed together in large towns; in this way it produces conditions that provoke an irreparable rift in the interdependence process of social metabolism, a metabolism prescribed by the natural laws of life itself. The result of this is a squandering of the vitality of the soil, which is carried by trade far beyond the bounds of a single country.

Large-scale industry and large-scale agriculture thus had the same results: both contributed to the ruining of the agricultural worker and to the exhaustion of the "natural power of the soil." "The moral of the tale," Marx observed, "is that the capitalist system runs counter to a rational agriculture, or that a rational agriculture is incompatible with the capitalist system (even if the latter promotes technical development in agriculture) and needs either small farmers working for themselves or the control of the associated producers."[28]

Unlike Malthus and the other classical political economists, Marx's tendency to seek out historical rather than "natural" laws took him far beyond the issues of population and soil fertility to larger issues of sustainability. For example, he argued that,"The development of civilization and industry in general has always shown itself so active in the destruction of forests that everything that has been done for their conservation and production is completely insignificant in comparison."[29] The capitalist mode of production, Marx wrote, "presupposes the domination of man over Nature." It treats Nature's contribution to productive wealth as a "gratuitous" gain or free gift.

---

**Engels described this freebooting relationship of humans to nature graphically: "The people who, in Mesopotamia, Greece, Asia Minor, and elsewhere destroyed the forests to obtain cultivable land, never dreamed that they were laying the basis for the present devastated condition of these countries, by removing along with the forests the collecting centres and reservoirs of moisture. When on the southern slopes of the mountains, the Italians of the Alps used up the pine forests so carefully cherished on the northern slopes, they had no inkling that by doing so they were cutting at the roots of the dairy industry in their region; they had still less inkling that they were thereby depriving their mountain springs of water for the greater part of the year, with the effect that these would be able to pour still more furious flood torrents on the plains during the rainy season.... Thus at every step we are reminded that we by no means rule over nature like a conqueror over a foreign people, like someone**

**standing outside nature—but that we, with flesh, blood, and brain, belong to nature, and exist in its midst, and that all our mastery of it consists in the fact that we have the advantage of all other beings of being able to know and correctly apply its laws."[30]**

---

## INDUSTRIALISM AND ROMANTIC ECOLOGY

From the beginning of the Industrial Revolution the conflict between machine capitalism, with its harsh and relentless impact on the environment, and an ecological ideal in which society and nature would no longer be alienated from one another was constantly invoked by Romantic social critics. In 1844, while Engels was writing *The Condition of the Working Class in England*, the Romantic poet William Wordsworth authored a sonnet protesting the building of a railway through the English lake district. "Is then no nook of English ground secure/From rash assault?" A year later, on the other side of the Atlantic, Henry David Thoreau retreated to Walden Pond outside Concord, Massachusetts, seeking through solitary communion with nature those principles that would offer an alternative to a world with which he was increasingly at odds. "I cannot believe," Thoreau wrote in the opening chapter of *Walden*, "that our factory system is the best mode by which men may get clothing. The condition of the operatives is becoming every day more like that of the English; and it cannot be wondered at, since, as far as I have heard or observed, the principal object is, not that mankind may be well and honestly clad, but unquestionably, that the corporations may be enriched."[31]

Walden Pond offered Thoreau no impenetrable retreat from the glaring contradictions of commercial and industrial society: it was connected to the social world by the train tracks that bordered the pond itself. His deep concern for the natural world was at one with his appreciation of craftsmanship, as opposed to modern industry; his opposition to the endless pursuit of wealth; and his rejection of the Southern slave system.[32]

Thoreau's commitment to both the natural world and to traditions of human craftsmanship, along with his critique of acquisitive society, had its counterpart in such English Romantic critics as John Ruskin and William Morris. Ruskin stressed the need for a more "organic society" influenced by the principles of art and intrinsic value, as opposed to mechanics and money. In *Unto This Last* (1860), he argued that production and possession may not constitute wealth but may instead be "illth," a word he coined. Wealth, he argued, was the "possession of useful articles, *which we can use.*" Conversely, the production of useless things, or the possession of things we cannot use, can only be defined as illth.[33]

Ruskin's ideas, along with those of Marx, helped inspire the great English artist, master-craftsperson, poet, and social critic William Morris, who merged Ruskin's Romantic critique of industrialism with a socialist cultural critique of capitalism. Morris believed, like Ruskin, that society should follow art. "Everything made by man's hands has a form," Morris wrote in 1878, "which must be either beautiful or ugly; beautiful if it is in accord with Nature, and helps her; ugly if it is discordant with Nature, and thwarts her." "Wealth," he later wrote, "is what Nature gives us and what a reasonable man can make out of the gifts of Nature for his reasonable use.... But think, I beseech you, of the product of England, the workshop of the world, and will you not be bewildered, as I am, at the thought of the mass of things which no sane man could desire, but which our useless toil makes—and sells?"[34]

For Morris, capitalist civilization offered a "stupendous organization—for the misery of life!" The method of distribution under capitalism, he wrote, "is full of waste; for it employs whole armies of clerks, travellers, shopmen, advertisers, and what not, merely for the sake of shifting money from one person's pocket to another's; and this waste in production and waste in distribution, added to the maintenance of the useless lives of the possessing and non-producing class, must all be paid for out of the products of the workers, and is a ceaseless burden to their lives." The real crime of modern civilization could not be traced to machinery or industrialization, but to the fact that in "using our control of the powers of Nature for the purpose of enslaving people, we care less meantime of how much happiness we rob their lives of."[35]

The root of this problem, for Morris, lay in "capitalistic manufacture, capitalistic land-owning and capitalistic exchange":

> It is profit which draws men into enormous unmanageable aggregations called towns, for instance; profit which crowds them up when they are there into quarters without gardens or open spaces; profit which won't take the most ordinary precautions against wrapping a whole district in a cloud of sulphurous smoke; which turns beautiful rivers into filthy sewers; which condemns all but the rich to live in houses idiotically cramped and confined at the best, and at the worst in houses for whose wretchedness there is no name.[36]

Morris's ecological critique did not stop with the urban environment. His broader commitment to environmentalism could be seen in his passionate defense of Epping Forest, near London. In opposition to the "expert" advice of "a wood bailiff, whose business is to grow timber for the market; or of a botanist whose business is to collect specimens for a botanical garden; or of a landscape gardener whose business is to vulgarize a garden or landscape to the utmost extent that his patron's purse will allow of," Morris, in a letter of protest to the *Daily Chronicle* in 1895, counterposed the ecological values of ordinary Londoners who had grown up with this unique forest and considered it part of the common wealth: "We want a thicket, not a park, from Epping forest.... [Otherwise] one of the greatest ornaments of London will disappear, and no one will have even a sample left to show what the great north-eastern forest was like."[37]

In the late nineteenth century the struggle over the conservation of nature was to emerge as one of the dominant forms of social struggle—and nowhere more so than in the United States.

# 4

# EXPANSION AND CONSERVATION

The conservation movement that arose in the late nineteenth century—most notably in the United States—was, according to noted natural-resource economists Harold Barnett and Chandler Morse, "an American part of a major revolution in thought throughout the Western world against the then-dominant social philosophy of the self-regulating market. Marxism was another European part of that same revolution in ideas." Just as socialists challenged the idea that labor should be looked upon as a mere "factor of production" in the operation of a competitive capitalist economy, so conservationists came to challenge the dominant notion of land as a mere economic factor. Only through the deliberate social regulation of land—often through public ownership—conservationists argued, could natural resources be preserved. A conflict emerged between private exploitation/short-term expansion and public conservation that was to be carried forward into the ecological disputes of the late twentieth century.[1]

## EXPANSION

The nineteenth century was a golden age of capitalist economic growth, and nowhere more so than in the United States. The second phase of the Industrial Revolution (1840-1875), which coincided with the spread of industrialization to the United States and Canada, was marked by the rise of the railroad and heavy industry. The building of railroads and canals opened up whole continents to the world market. By the 1850s the railroad alone accounted for around 15 percent of all investment in the United States. By the end of the century, however, the great railroad boom was over and both economic growth and geographic expansion began to slow. In 1890 the Bureau of the Census officially declared the "frontier of settlement" closed.[2]

Three years later, in a talk before the American Historical Association in Chicago, Frederick Jackson Turner introduced his famous "frontier thesis." According to Turner, the history of the United States up to 1890 could be viewed as a great open book in which the central theme was the conquest of successive frontiers:

> It begins with the Indian and the hunter; it goes on to tell of the disintegration of savagery by the entrance of the trader, the path-finder of civilization; we read the annals of the pastoral stage in ranch life; the exploitation of the soil by the raising of unrotated crops of corn and wheat in sparsely settled farming communities; the intensive culture of the denser farm settlement; and finally the manufacturing organization with city and factory system.[3]

Each new natural environment, in Turner's view, created a different type of frontiersman: "The exploitation of the beasts took hunter and trader to the west, the exploitation of the grasses took the rancher west, and the exploitation of the virgin soil of the river valleys and prairies attracted the farmer." Each frontier "was won by a series of Indian wars" that conquered both "hostile

Indians and the stubborn wilderness." In this respect the frontier was "a military training school, keeping alive the power of resistance to aggression, and developing the stalwart and rugged qualities of the frontiersman."[4]

For Turner, the economic prosperity and democracy of the United States was rooted in the availability of "free land." The closing of the frontier meant that there was no more free land to be exploited. Both democracy and propsperity were therefore threatened. The only answer was to seek new frontiers abroad.

Such expansionist views of were not without their critics. In the eyes of the great social critic Thorstein Veblen, the frontier expansion of American society—the later stages of which were carried out not so much by Turner's frontiersman as by large, monopolistic corporations—was a story of waste on a gigantic scale. Veblen believed that business civilization had created a culture in which predatory and pecuniary values dominated over values of craftsmanship, industry, and conservation. Presenting a view of the frontier quite different from Turner's, he argued in *Absentee Ownership* (1923) that the "American plan or policy" of natural resource use "is very simply a settled practice of converting all public wealth to private gain on a plan of legalized seizure." The first natural resources to be seized and exhausted in this way had been the fur-bearing animals. This same practice of enrichment through the systematic looting of natural wealth had then been applied to the soil in the slave South, which was ruined by cotton production; to mining for precious metals; and to the exploitation of "timber, coal, iron and other useful metals, petroleum, natural gas, water-power, irrigation [resources], [and] transportation (as water-front, right-of-way, terminal facilities)." "Capitalizing" on natural resources by treating them as a source of "free income," Veblen argued, encouraged waste on a national and global scale. For instance, the wasted timber associated with logging and land-clearing practices was so great "that this enterprise of the lumber-men during the period since the middle of the nineteenth century has destroyed very appreciably more timber than it has utilised."[5]

## CONSERVATION

For most conservationists concerned with the preserving the wilderness, it was not simply the frontier that was vanishing in the face of the human onslaught, but that part of nature that was independent of human beings. When Vermont naturalist George Perkins Marsh wrote his classic *Man and Nature*—referred to by Lewis Mumford as "the fountainhead of the conservation movement"—in 1864, it was the most detailed and systematic study of "the earth as transformed by human action" since Buffon's work of the late eighteenth century (see Chapter 2). However, Marsh's views were a far cry from Buffon's optimistic Enlightenment appraisal of the ability of humans to dominate nature. "There are parts of Asia Minor, of Northern Africa, of Greece, and even of Alpine Europe," Marsh wrote,

> where the operation of causes set in action by man has brought the face of the earth to a desolation almost as complete as that of the moon; and though, within that brief space of time which we call "the historical period," they are known to have been covered with luxuriant woods, verdant pastures, and fertile meadows, they are now too far deteriorated to be reclaimable by man, nor can they become again fitted for human use, except through great geological changes, or other mysterious influences or agencies of which we have no present knowledge, and over which we have no prospective control. The earth is fast becoming an unfit home for its noblest inhabitant.... Another era of equal human crime and improvidence ... would reduce it to such a condition of impoverished productiveness, of shattered surface, of climatic excess, as to threaten the depravation, barbarism, and perhaps even extinction of the species.[6]

No one placed greater emphasis than Marsh on the fact that "nature" was not something external to human beings, but that human beings were themselves authors of nature. *Man and*

*Nature*, as he wrote to a friend before the book was published, was intended as "a little volume showing that whereas [others] think that the earth made man, man in fact made the earth." In other words, nature could no longer be seen as a reality external to human society, but was to a considerable degree the product of human transformation. Marsh's approach therefore brought the issue of nature into human history as never before. Moreover, humans were transforming the earth in alarming ways. After carrying out an audit of the entire planet, Marsh concluded that, "Man is everywhere a disturbing agent. Wherever he plants his foot, the harmonies of nature are turned to discords.... [O]f all organic beings, man alone is to be regarded as essentially a destructive power, and ... wields energies to resist which, nature ... is wholly impotent."[7]

Although industrialization entered only indirectly into Marsh's analysis, it was clear that it constituted the major force behind the ecological destruction that he described. It is thus no mere coincidence then that *Man and Nature*—the leading work on planetary ecological devastation to appear prior to the twentieth century—was published in 1864, only three years before *Capital*, Karl Marx's famous critique of the age of industrial capital. Both works were a response to forces engendered by the Industrial Revolution. And while Marx's ideas helped to inspire working-class revolts against capitalism, Marsh's ideas gave impetus to the struggle to place limits on the human exploitation of nature.[8]

The year 1864, when *Man and Nature* was published, marked a turning point for public conservation: the U.S. government ceded Yosemite Park to the state of California with the stipulation that it be preserved as a public park. A few years later, in 1872, the federal government established Yellowstone as the first National Park. The conservation movement that emerged from these modest beginnings is usually identified with such great figures as Marsh, Frederick Law Olmsted, John Muir, and Aldo Leopold, along with more business-oriented conservationists such as Gifford Pinchot and Theodore Roosevelt. Yet, it is worth remembering that many other individuals, most now forgotten, protested the wanton destruction of nature and the extreme environmental degradation of the cities. A large proportion of these protesters were women. For instance, about half of the nature essays written for the *Atlantic Monthly*, when this became a recognized genre in the late nineteenth century, were by women. In the very first issue of *Audubon Magazine*, in 1887, Celia Thaxter wrote against the feminine fashion of decorating hats with the feathers of rare birds.[9]

A major concern, leading to the growth of the conservation movement, was the rate of extinction. By the late nineteenth century it was clear, to those who saw the transformations of the U.S. wilderness through conservationist eyes, that the extermination of wildlife was occurring everywhere. Some 40 million bison (or buffalo) had ranged over a third of North America when the Europeans first arrived. Commercial hunting of bison for meat began in the 1830s and soon reached 2 million head a year; it rose to 3 million a year after 1870, when bison hides began to be made into commercial leather. The Union Pacific Railroad, completed in 1869, divided the bison into northern and southern herds, which made them easier to hunt. The southern herd was largely exterminated in the early 1870s. After the completion of the Northern Pacific Railroad in 1880, the slaughter of the northern herd commenced. By the last decade of the nineteenth century, bison were nearly extinct. "The bison ... met their end," environmental historian William Cronon has written, "because their ecosystem had become attached to an urban marketplace in a new way."[10]

As bird species vanished, bison herds disappeared, and forests became mere memories, more and more people, particularly in the growing urban centers of the country, became concerned about conservation. The great enemies of nature in the popular view were the land-grabbing railroads and the large logging companies. The conservation movement thus received much of

**Men on the transcontinental railroad shooting bison for sport. [Culver Pictures, Inc.]**

its impetus from populist attacks on railroads, Gilded Age capital, and big business. Despite this, the conservation movement came to be dominated by more business-oriented forces who sought not so much to oppose the environmental depredations of the large corporations as to regulate and rationalize the exploitation of natural resources for purposes of long-term profits. Hence the figures who came to exemplify the dominant strand of U.S. conservationism during this period were not preservationists like John Muir, the defender of the Yosemite and the founder of the Sierra Club, but self-styled "scientific-managers" like Gifford

Pinchot and Theodore Roosevelt, the founder of the U.S. Forest Service, who advocated "efficiency," "wise use," and the application of business principles to nature.[11] It was the efficiency-minded conservationists who were most effective in exerting direct control over state policy in order to construct a system of natural resource management based in government agencies.

Thus Pinchot, the founder of the U.S. Forest Service and the preeminent figure in the creation of the national forests, made a sharp distinction between "scientific forestry" and preservation—a distinction symbolized by his successful effort to have the national forest reserves placed under the administration of the Department of Agriculture and to have the name changed from "forest reserves" to "national forests." "The object of our policy," Pinchot explained, "is not to preserve the forests because they are beautiful ... or because they are refuges for the wild creatures of the wilderness ... but ... the making of prosperous homes.... Every other consideration comes as secondary."[12]

But despite the narrow economic motives that guided Pinchot's conservationism, it represented an attempt to move away from pure market principles where natural resources were concerned. Progressive conservationists thus stood for what Pinchot's advisor Philip Wells was to refer to as the "socialization of management" in the use of natural resources.[13]

President Theodore Roosevelt's approach to conservation was a complex amalgam of business spokesperson concerned with long-term profits, engineer concerned with the most efficient exploitation of natural resources, and hunter loath to see the last large game animals disappear. As he told the Forest Congress in 1905, "Both the production of the great staples upon which our prosperity depends and their movement in commerce throughout the United States are inseparably dependent upon the existence of permanent and suitable supplies from the forest at reasonable cost." Despite the obvious limitations associated with such a stance, Roosevelt's conservationism produced positive results: for instance tens of millions of acres of forest were added to the National Forests during his years as President (1901 to 1909).[14]

In contrast to this dominant form of conservationism, there were others who represented a wider "ecological conscience" rooted in a nascent "land ethic"—to use terms coined later by ecologist Aldo Leopold. In Leopold's words, "The land ethic simply enlarges the boundaries of the community to include soils, waters, plants, and animals, or collectively: the land." It rejects the valuation of the biotic (or natural) world in economic terms alone and establishes distinct "biotic rights" to existence. As the U.S. conservation movement developed, it was more and more divided between those establishment figures like Pinchot and Roosevelt who saw conservation in purely monetary terms—that is, in terms of the "wise use" of natural resources—and those like Muir and Leopold who insisted on putting preservation first and profits second. The most famous struggle between the two branches of the movement was to occur in the years 1908-1913, when Muir and his allies (including the major women's clubs attached to the conservation movement) attempted to prevent the beautiful Hetch Hetchy Valley in Yosemite National Park from being turned into a reservoir that would provide hydroelectric power for San Francisco. In the end, Pinchot and Roosevelt won and the dam was built.[15]

If the main struggle within the conservation movement from the 1890s to the 1930s was between preservationists, who emphasized the intrinsic value of nature, and conservationists, who worshiped at the altar of efficiency, there were others who did not fit comfortably into either of these camps. The conservation movement also attracted political radicals, such as the socialist preservationist Robert Marshall (1901-1939). Marshall, an employee of the Forest Service, which he served in various high-level capacities, was one of the leading figures involved in the struggle to reform the Forest Service's management of the national forests during the New Deal Presidency of Franklin Roosevelt. He was also one of the founders of the Wilderness Society (in 1935), and its principal source of funds in its early years. Marshall believed that it was essential to preserve vast tracts of wilderness as roadless areas free from mechanical contrivances of every kind. At the same time he was the strongest critic in his day of the private exploitation of forests. In his best-known work, *The People's Forests* (1933), he argued that "The fundamental advantage of public ownership is that in the former social welfare is substituted over private gain as the major objective of management." It was therefore necessary to protect wilderness areas from "commercial exploitation" by expanding public ownership. He recommended that the government immediately acquire an additional 240 million acres of forest. He had a number of plans for how this land could be used: it could become a center of employment for several million of the 12 million unemployed. He also wanted government subsidized public transportation leading to the public forests; creation of camps for urban workers to enjoy the forests at nominal cost; the elimination of Forest Service practices that discriminated against blacks, Jews, and other minorities; and purchase of more recreational public land near urban centers. In response to his promotion of such ideas, the editor of *The Journal of Forestry* labeled him a "dangerous radical." In 1938, the year before his death, Marshall came under investigation by the House Committee on Un-American Activities, which accused him of aiding Communism through his support of radical causes. In his will Marshall divided his $1.5 million estate equally among three trusts: the first devoted to the promotion of trade unions and "an economic system in the United States based on the theory of production for use and not for profit"; the second backing civil liberties; and the third committed to wilderness preservation.[16]

## THE CITIES

While the conservation movement was struggling to define the future of what remained of the American wilderness, there were other environmental struggles taking place in the "urban wilderness" or "jungle." These late-nineteenth and early-twentieth century urban environmental struggles were organized around the demand for improved public health and sanitation. Even though it had been understood since the 1830s and 1840s that many of the worst epidemics could be countered through improved water and sewage facilities, progress had been slow. As late as 1866 a cholera epidemic broke out in London, killing 6,000. The infant mortality rate in New York City was 240 per thousand live births in 1870. Yet the provision of sanitary facilities was delayed because private firms did not consider such facilities remunerative. As a result, local communities everywhere were forced to turn to what English radical Sidney Webb was to call "municipal socialism" as a means of establishing the necessary environmental safeguards.[17]

---

**"Just as early industrialism had squeezed its profits not merely out of the machine, but out of the pauperism of the workers," Lewis Mumford has written, "so the crude factory town had maintained its low wages and taxes by depleting and pauperizing the environment. Hygiene demanded space and municipal equipment and natural resources that had hitherto been lacking. In time, this demand forced municipal socialization, as a normal accompaniment to improved service. Neither a pure water supply, nor the collective disposal of garbage, waste, and sewage, could be left to the private conscience or attended to only if they could be provided for at a profit.... In small centers, private companies might be left with the privilege of maintaining one or more of these services, until some notorious outbreak of disease dictated public control; but in the bigger cities socialization was the price of safety; and so, despite the theoretic claims of *laissez faire,* the nineteenth century became, as Beatrice and Sid-**

> **ney Webb correctly pointed out, the century of municipal socialism. Each individual improvement within the building demanded its collectively owned and operated utility: watermains, water reservoirs and aqueducts, pumping stations: sewage mains, sewage reduction plants, sewage farms.[18]**

The goal of bringing fresh air, clean water, green open spaces, sunlight, and fresh food back into the city, while combatting poverty, slums, and homelessness, brought together planners, physicians, public health officials, members of women's clubs, romantic critics of capitalist society, and socialists, all of whom sought to make the degraded urban environment liveable. In the United States, the centrality of the women's movement to the struggle for the urban environment led to an emphasis on "municipal housekeeping" that united the battles over the home and urban environments into one struggle for reform. Some of the most prominent women in the country were leaders in this struggle, including Jane Addams, founder of the legendary Hull House in Chicago (a settlement house in the working-class quarter of the city occupied by women dedicated to the cause of social reform); Florence Kelley, a translator and friend of Frederick Engels and the most indefatigable Hull House researcher into the conditions of sweatshops located in tenement houses; Alice Hamilton, the first woman professor at Harvard and the pioneering figure in occupational health in the United States; and Ellen Swallow Richards, the first woman to attend MIT, who went on to play a leading role as a scientist and reformer in the areas of sanitation and nutrition.[19]

In England the conditions of the cities at the turn of the century inspired the utopian socialist planner Ebenezer Howard to develop his "garden city" (or green belt) plan for naturalizing the city environment and transcending the contradiction between city and country. Howard's ideas were to exert a strong influence

**An overcrowded tenement on New York's Lower East Side in the early 1900s. [Culver Pictures, Inc.]**

on later radical urban and regional environmentalists in both Britain and the United States, including Sir Patrick Geddes and Lewis Mumford.[20]

Socialists and radical planners were often leaders in this movement, since they saw the need for public as opposed to private solutions and were attuned to working-class demands for change. Indeed, the most insistent calls to reform the city environment often came from the bottom of the social hierarchy. For instance, it was the organized protest of thousands of unemployed workers in New York City in 1857, calling day after day for the construction of Central Park (along with other public work projects) that led to the building of the park under the direction of the famous landscape architect Frederick Law Olmsted.[21]

Building on this popular desire for the improvement of the urban environment, socialists called for the reorganization of both society and urban life. Edward Bellamy's utopian socialist novel *Looking Backward* (1888), which sold millions of copies in the United States, owed much of its popularity to the sharp contrast it drew between the environment of nineteenth-century capitalist Boston and a utopian socialist Boston of the year 2000. The picture the novel painted of Boston in the year 2000 was idyllic: "At my feet lay a great city. Miles of broad streets, shaded by trees and lined with fine buildings, for the most part in continuous blocks but set in larger or smaller inclosures stretched in every direction. Every quarter contained large open squares filled with trees, among which statues glistened and fountains flashed in the afternoon sun."[22] City and regional planners such as Boston's Sylvester Baxter saw Bellamy's vision of the city of the year 2000 as indistinguishable from the efforts of those like Olmsted who sought to build a system of city parks across the nation, urbanizing nature in order to naturalize the urban environment. Bellamy's ideas caught fire in this atmosphere of urban reform because they envisioned a city without poverty, crime, or violence, one in which society's economic surplus was used primarily for public works and pleasures rather than for the increase of individual incomes, and where the splendor of life was on its social side. He believed that a rationally organized system of production could meet all essential human needs, and depicted a society in which people were not stimulated to consume more and more.[23]

For another influential socialist writer, Upton Sinclair, the capitalist urban environment was a "jungle"—a metaphor that he employed in his great muckraking novel by that title (originally serialized, beginning in 1904, in the socialist publication *The Appeal to Reason*). Sinclair's *The Jungle* had as its subject Chicago's Packingtown, best known for its immense slaughterhouses. Many of its most powerful passages were devoted to environmental destruction. In describing the part of the Chicago River known as "Bubbly Creek," he wrote:

> It is constantly in motion, as if huge fish were feeding in it, or great leviathans disporting themselves in its depths; bubbles will rise to the surface and burst, and make rings two or three feet wide. Here and there the grease and filth have caked solid, and the creek looks like a bed of lava; chickens walk about on it, feeding, and many times an unwary stranger has started to stroll across, and vanished and never been seen again. The packers used to leave it that way, till every now and then the surface would catch fire and burn furiously, and the fire department would have to come and put it out.[24]

Chicago, with its immense slaughterhouses, was the point at which the railways of the West converged with those of the East. Because of this it was the most important urban center for the exploitation of the natural resources of the Western hinterland. Cattle were born in the West, grazed on lands formerly occupied by the buffalo, fattened on their way to market in the feedlots of Iowa and Illinois, and slaughtered on the "disassembly line" in Chicago. There they were converted into final commodities in the form of dressed beef to be shipped East. The object of this corporate meat-packing was, in William Cronon's words, "to systematize the market in animal flesh—to liberate it from nature and geography." For Upton Sinclair this was nothing other than "the spirit of Capitalism made flesh."[25]

---

**In a famous description of Chicago's slaughterhouses in his novel *The Jungle* (1904), Upton Sinclair emphasized the common degradation to which both animals and human beings were reduced under capitalism: "It was pork-making by machinery, pork-making reduced to mathematics. And yet somehow the most matter-of-fact person could not help thinking of the pigs; they were so innocent, they came so very trustingly; and they were so**

**very human in their protests—and so perfectly within their rights! They had done nothing to deserve it; and it was adding insult to injury as the thing was done here—swinging them up in this cold-blooded, impersonal way, without a pretense of apology, without the homage of a tear. Now and then a visitor wept, to be sure; but this slaughtering-machine ran on, visitors or no visitors.... Who would take this pig into his arms and comfort him, reward him for his work well done, and show him the meaning of his sacrifice? For all the while there was a meaning—if only the poor pig could have known it. Perhaps if he had, he would not have squealed at all, but died happy! If only he had known that he was to figure in the bank-account of some great captain of industry, and perhaps help to found a university, or endow a handful of libraries, when the captain of industry died! It is one of the crimes of commercialism that it thus cruelly leaves its victims to grope in darkness; that delicate women and little children, who toil and groan in factories and mines and sweatshops and die of starvation and loathsome diseases, are not taught and consoled by the reflection that they are adding to the wealth of society, and to the power and greatness of some eminent philanthropist."[26]**

---

This spirit visible in thc rclationship between Chicago and its hinterland was also increasingly evident on a world scale in the larger division between global center and periphery. Here too the environment was being systematically "liberated" from both nature and geography in the endless search for economic gain. We turn to this now.

# 5

# IMPERIALISM AND ECOLOGY

From its very earliest beginnings in the late fifteenth and early sixteenth centuries, capitalism has always been a world system, dividing the globe into center and periphery. The existence of such a hierarchy has meant that the people and the ecosystems of the periphery have been treated as appendages to the growth requirements of the advanced capitalist center. Each stage of capitalist development—mercantilism, early industrial capitalism and monopoly capitalism—has seen the expansion of this imperialist relation to the planet.

## COLONIALISM AND ECOLOGY

Parts of the Americas were fully integrated as dependent peripheries within the world economy by the seventeenth century, but the integration of Asia and Africa occurred later. When European sea merchants first began to expand into Asia in the sixteenth century, the land mass was still under the control of powerful tributary states—states that were larger, more densely populated, and frequently more productive than European capitalist states of the period. Superior sea power, however, allowed the Europeans to dominate the oceans and establish footholds on the "tidal margins" in certain key areas. In the sixteenth century, the leading European power to expand into Asia was Portugal, which during the seventeenth century was increasingly displaced by the Netherlands and England.[1]

The English concentrated their activities on the Indian subcontinent, where, beginning in 1611, they established what were called "factories"—points of settlement and commerce—along the coast. Eventually the British East India Company (a Crown-licensed monopolistic trading company) went to war against the local rulers in Bengal, defeating them in the battle of Plassey in 1757.

Following this, the Company rapidly expanded its territorial domain, largely by means of successive wars, taking over some areas directly while leaving others to the administration of local rulers (under the Company's sponsorship). After 1814, the high-quality textile crafts of the Indian towns collapsed when British machine-manufactured cotton textiles flooded the market. Meanwhile, Indian property holders began to produce cash crops for export to Europe. In the case of British India, however, this occurred not through the creation of single-crop plantations but through the formation of an agricultural putting-out system in which crops were purchased by intermediaries, who passed them on to the final buyer. In this situation, moneylenders increasingly preyed upon hard-pressed peasant producers. The imposition of heavy land taxes under British rule paralyzed agriculture, preventing its development. At the same time, the economic surplus siphoned from India helped feed British industrialization. In this way, India was gradually integrated into the capitalist world economy as a dependency of Britain, the most important possession in the British empire.[2]

The conquest of India eventually created the conditions for the British penetration of China. Desiring tea and other goods, the British organized the production of opium in Bengal in order to obtain a commodity that would finally open the door to extensive trade in China. This was to lead to the Opium Wars in China, as the Chinese resisted the systematic imposition of this addictive drug on their society. The defeat of the Chinese Empire in the first Opium War (1839-1842) signalled that all non-Western societies had become helpless in the face of Western military and economic aggression.[3]

By the end of the nineteenth century, prospects for further expansion were limited because most of the earth had been parcelled out to one metropolitan power or another. Between 1876 and 1915, around one-quarter of the land surface of the globe was formally annexed and distributed as colonies to half a dozen states. Except for Ethiopia, Liberia, and part of Morocco, Africa was divided between Britain, France, Germany, Belgium, Portugal, and Spain. Although most of the great traditional empires of Asia remained nominally independent, the leading capitalist powers had carved out "zones of influence" and subjected these to unequal treaties.[4] In Latin America, most states, though also nominally independent, had been reduced to being British and U.S. economic dependencies.

The spirit of imperialism was best conveyed by the British statesman Cecil Rhodes, founder of Rhodesia, who is reported to have stated: "I would annex the planets if I could." And there was never any doubt about the ultimate means of such annexation. In the immortal lines of poet Hilaire Belloc, "Whatever happens, we have got/The Maxim Gun, and they have not.[5]

Rhodes explained the motivation behind British imperialism in this way: "We must find new lands from which we can easily obtain raw materials and at the same time exploit the cheap slave labour that is available from the natives of the colonies. The colonies would also provide a dumping ground for the surplus goods produced in our factories."[6] European economic historians dubbed the entire period from the mid-1870s to the mid-1890s as the "Great Depression." Imperialism was the most widely advocated remedy for this condition, both in Europe and the United States.

For Frederick Jackson Turner and Theodore Roosevelt, such expansion was simply an extension of the frontier. As Turner wrote in the *Atlantic Monthly* in 1896, "For nearly three centuries the dominant fact of American life has been expansion.... The demands for a vigorous foreign policy, for an interoceanic canal, for a revival of our power upon the seas, and for the extension of American influence to outlying islands and adjoining countries, are indications that the movement will continue."[7] In April 1898 the United States launched the Spanish-American War. Within three months the Spanish had been defeated and the United States had laid claim to an informal empire consisting of Cuba, Puerto Rico, Guam, and the Philippines. In 1898 the United States annexed Hawaii, and in 1899 the Samoan Islands were partitioned between the United States and Germany.

Little attempt was made by the dominant business interests to disguise the fact that the main objectives of this expansion were economic. In the midst of the war, the *Lumberman's Review,* a key organ of the lumber industry, declared: "The moment Spain drops the reigns of government in Cuba ... the moment will arrive for American lumber interests to move into the island for the products of Cuban forests. Cuba still possesses 10,000,000 acres of virgin forest abounding in valuable timber ... nearly every foot of which would be saleable in the United States and bring high prices."[8]

Colonialism and imperialism pillaged the ecologies and societies of the conquered territories, while contributing relatively little to their economic progress. In the periphery of the world economy, the Industrial Revolution led not to development but to what Andre Gunder Frank has called "the development of underdevelopment."[9]

This can be seen most clearly in the case of the railways. Of all the nations that built large railway lines in the nineteenth century—in order of magnitude (in 1900), the United States, Russia, Germany, British India, France, Britain, and Canada—only British India failed to industrialize during the railway boom. The reason was that the goal of railway building in India was not to develop *India* but to develop Britain. The cotton manufacturers of Lancashire saw the Indian railway as a mere extension of the line from Manchester to the port of Liverpool. In addition, the Indian Mutiny of 1857 had demonstrated the strategic value of railways in moving troops and equipment rapidly. Railways were therefore built to serve the economic and military objectives of British colonialism. Although 4 percent of the locomotives used in British India were built there, British imperial policy ensured that 80 percent were imported from Britain, while the remaining 16 percent were imported from Germany and the United States. As a result of this policy, India was the only country with extensive railways that did not also develop a strong locomotive-building industry.

India's fate in this respect contrasts sharply with that of Japan, which was exceptional in the sense that it was never colonized. There the number of foreign railway technicians rose from 19 in 1870 to 113 in 1874 and then dropped to 43 in 1879 and 15 in 1885—after which the Japanese dispensed with foreign technicians altogether. Unlike those of India, Japan's railways were directed to the needs of the domestic economy, not to the interests of a colonial power.[10]

Colonization in the industrial age meant that the division between town and country was extended into the periphery of the world economy. Western urban technologies of water supply and sewage disposal were also imported, but while in the advanced capitalist countries there was generally a lag before the benefits of these new facilities reached the poor, in the colonial world they frequently never did. As historian Daniel Headrick has written: "Colonial cities gathered increasing numbers of people with little access to those sanitary improvements which European city planners and administrators were so proud to have introduced.... In the Western world after the mid-nineteenth century, municipal health officials and sanitation engineers strove to separate the germs from the people. In tropical cities, when the officials could not achieve this objective, they substituted another: to separate the people with germs from those without."[11]

In Dakar, Senegal, in the early twentieth century the segregation of the city into a European town and a native town was justified by the colonial authorities in terms of the types of buildings and hygienic habits that characterized the distinct lifestyles of the Europeans and non-Europeans. As the municipal governor of Senegal wrote to the municipal council, "Let us allow them, if need be let us make them, have two different installations conforming to their tastes: on the one side the European town with all the requirements of modern hygiene, on the other the native town with all the freedom to build out of wood or straw, to play the tom-toms all night, and to pound millet from four in the morning on." The inequalities in the distribution of wealth, availability of accommodation, and provision of water and sewage facilities that lay behind these differences in "taste" and hygiene were simply ignored—as if they were all a product of different "lifestyles."[12]

The international division of labor that concentrated industrialization in the advanced capitalist countries and the production of agricultural export crops and extractive industries in the periphery contributed to the rapid rate of ecological degradation in the periphery. In places as varied as Brazil, Egypt, Gambia, the Gold Coast, and Senegal, export agriculture severely eroded and depleted the soil. Throughout the periphery, the neglect of subsistence food production resulted in the death from famine and nutritional deficiencies of untold numbers of people. In contrast, by the 1890s Japan was able to enter the ranks of the advanced capitalist states and was largely free of such problems. Up to the 1950s, in fact, it was possible to contend that "for all practical purposes, there is no such thing as erosion in Japan."[13]

## ECOLOGICAL IMPERIALISM

The extent to which the ecology of the periphery was transformed to meet the requirements of the center can be seen dramatically in the way in which European botanical science was organized so as to exploit tropical agriculture. From the beginning of colonization, Europeans had craved tropical products: at first sugar, coffee, and tea, and later cotton, quinine, and rubber. As Daniel Headrick has observed, "Plants are the wealth of the tropical world and the livelihood of most of its people."[14] But it was not until the Industrial Revolution that global plant transfers began to be systematized.

The key institutions in this process were the botanical gardens. The Netherlands, Britain, and France all used a system of botanical gardens to control plant transfers across their empires. The most important of these was Kew Gardens, the royal botanical garden outside London, which was founded in 1772 and soon emerged as a center for botanical research. In the eighteenth century, plant transport was difficult, since living plants tended to perish in the long journey across the seas. For this reason the early botanical gardens mainly collected and classified information about tropical plants. In 1829, however, Dr. Nathaniel Ward invented a large terrarium—a wooden box with a glass top—that kept delicate plants from drying out and allowed them to be transported over great distances. The "Wardian case" initiated a new era of plant transfer and Kew became the central institution in a hierarchy of botanical gardens that was established throughout the British empire.

Beginning in the 1840s, botanists at Kew began to work systematically on plant transfers. In 1872, Liberian coffee was first transferred from West Africa to areas throughout the world; in 1876, the rubber tree was transferred from Brazil to Ceylon. The ultimate goal was to enhance production for the market by bringing together easily commodifiable plant species and the most cost-efficient supplies of human labor. As a result of the efforts of collectors who sent in seeds and plants from around the world, Kew had over 1 million plants in its gardens and herbarium by the close of the century, and could identify more plant species than any other institution in the world. Kew supervised satellite institutions in places like Hong Kong, Tasmania, Natal, Calcutta, and Singapore. "At very little cost to Britain," explains Lucille Brockway, author of *Science and Colonial Expansion*, "Kew searched for economically useful plants, most often in South America, ... improved them, and transferred them to Asia, thus participating in extending the plantation system to Asia."[15]

An example of the imperialist role played by the botanical gardens can be seen in the case of rubber. The most highly prized of the rubber plants was *Hevea brasiliensis*, which was native to the Amazon rain forest. In 1876 a British planter, Henry Wickham, smuggled 70,000 Hevea seeds out of the Amazon. These seeds made their way to Kew, where they were germinated. The seedlings were then sent to locations throughout the British empire. For three years a large part of the British imperial botanical establishment was mobilized to ensure the success of these rubber transfers. The most important occurred when twenty-two seedlings were sent from Ceylon to Singapore in 1877. From these arose almost all of the rubber trees now to be found in Southeast Asia.

The transfer and transformation of species within a world system of commodity exchange replaced natural complexity with the simplicity of commodified agriculture. Just as the Industrial Revolution made possible the subjection of labor to capital, so it also made possible the subjection of nature to capital. The division of labor, which formed the basis of accumulation, was accompanied by the division of nature. The end result of this

**Rubber companies used conscripted workers to raze the jungle. This was a Goodyear publicity photo. [Culver Pictures, Inc.]**

process, which was to be extended enormously in the twentieth century, was "genetic erosion": the loss of diversity. Not only were invaluable species lost through the extension of agricultural monocultures but crucial varieties (landraces built up over thousands of years) of key crops disappeared as well. Increased reliance on a few genetic varieties of cultivated crops made these crops more vulnerable to natural hazards. For instance, in 1970 the great southern corn leaf blight destroyed 15 to 20 percent of the total U.S. corn crop.[16]

This loss of diversity can be understood more fully in the

context of what has come to be known as the "Green Revolution" in twentieth-century agriculture: the energy-intensive application of synthetic fertilizers, pesticides, certain "miracle" seed varieties, and petroleum-powered machinery, substituting these for traditional agricultural methods. At the heart of this process has been the development of certain varieties of seeds that fit the Green Revolution technology.

Human beings have historically cultivated more than 3,000 plant species for food. Today fifteen species—among them rice, corn, wheat, potato, cassava, the common bean, soybean, peanut, coconut, and banana—provide 85 to 90 percent of all human energy. Three of these—rice, corn, and wheat—supply 66 percent of the world's entire seed crop. In 1920, the Russian plant geneticist N.I. Vavilov determined that there were a number of centers of great plant gene diversity, all located in the underdeveloped countries (along the Tropic of Cancer and the Tropic of Capricorn), particularly in isolated mountainous areas. For many years scientists have been returning to these genetic "reservoirs" (in places like Mexico, Peru, Ethiopia, Turkey, and Tibet) for new germplasm to use in breeding resistance in commercial varieties. Yet these genetic reservoirs are not being adequately protected, and many of their genetic materials were discarded by scientists when they did not appear immediately useful. And as Western-dominated monocultural agriculture invades these regions of extreme diversity, they are being converted into areas of seed uniformity. "With the coming of plant and gene patenting and the opportunity of monopolization," world-famous authorities on plant genetic resources Cary Fowler and Pat Mooney write, "international companies have attempted to corner the market for the vanishing genes. The result may be the shattering of agriculture itself." In the last two decades, around 1,000 traditional seed companies have been absorbed by international biochemical corporations. The old system of botanical gardens, through which the developed countries coordinated the transfer of valuable tropical plant species, has now been replaced by a system of International Agricultural Resource Centers (IARCs), which are part of the Consultative Group on International Agricultural Research (CGIAR) system headquartered at the World Bank. The IARCs serve as mechanisms that transfer the plant genetic resources of the third world to the gene banks of the advanced capitalist states.[17]

Economic and ecological imperialism in this area is severe. CIMMYT (the International Center for Maize and Wheat Improvement) in Mexico provides countries with improved breeding material from its genetic resources. Participating countries are allowed to keep the germplasm. The U.S. Agency for International Development (AID) provides CIMMYT with an annual of $6 million for its work on cereals. In return, the United States receives ready access to Mexican germplasm. In 1984 alone, the U.S. government estimated that CIMMYT material contributed almost $2 billion to the value of the wheat crop as it left the farm. Meanwhile, lacking funds for its research, CIMMYT has turned over much of its exotic material for corn breeding to Pioneer Hi-Bred, the world's largest corn-breeding corporation. In a development that is viewed with alarm in the third world, IARCs have recently provided germplasm to multinational corporations, which have then patented the germplasm (in thinly disguised forms) in the United States and the United Kingdom. It is no wonder, then, that in the Philippines the seeds of the International Rice Research Institute (which is located in the Philippines but is controlled by international capital) have been called "the seeds of imperialism." As Robert Onate, president of the Philippines Agricultural Economics and Development Association, has stated, the "Green Revolution connection" can be understood as: "New seeds from the ... global crop/seed systems which will depend on the fertilizers, agrichemicals, and machineries produced by conglomerates of transnational corporations." The overall result is therefore to increase both the economic dependency and ecological vulnerability of third world nations. Such dependency and vulnerability arising from the Green Revolution was brought into sharp relief in Bhopal, India in 1984, when a leakage from a Union Carbide pesticide plant poisoned more than 200,000 people, killing some 3,000.[18]

## THE ENVIRONMENT OF THE COLD WAR: ECOCIDE IN THE SOVIET UNION

For most of the twentieth century, the world was divided not only between core and periphery, or North and South, but also between East and West. Consequently, imperialism in this period cannot be understood without also taking the Cold War into consideration. From the moment of its emergence in the October Revolution of 1917—occurring in the midst of World War I—the Soviet Union was faced with the enmity of the advanced capitalist world. It was not until after World War II, however, that this conflict took center stage in world history. Under U.S. leadership the West sought to isolate the U.S.S.R. and to push back revolutionary forces throughout the third world. Both the United States and the Soviet Union thus became involved in a global arms race that consumed vast economic and environmental resources.

Surrounded by a much larger, more powerful capitalist world economy, from the moment its society began to take definite shape in the late 1920s the Soviet Union was a classic example of a war economy: the forced drafting of labor, natural resources, and productive capacity and the building of a strong military defense complex were to characterize its entire development. These goals were implemented by a hierarchical command system—far removed from the genuine socialist ideal of social governance of production—which became entrenched as a result of the Stalinist purges of the mid-1930s, and which dominated not only the military and economy but also most aspects of civilian life. Compelled by the arms race to carry a military burden big enough to counterbalance that of the larger advanced capitalist system, the Soviet Union was forced to devote a large percentage of its total output—as much as 13 percent of GNP in the 1970s according to the CIA—to national defense. Equally disastrous for its own development, the Soviet Union was from the first dependent on a complex of technologies that had been developed in the capitalist world, as well as class-oriented techniques of labor organization, such as scientific management or Taylorism, developed in the West. Ecologically, the Soviet Union pioneered in various forms of environmental reform. But with the rise of the Stalinist bureaucracy in the 1930s, all of this gave way before the single goal of expanding production. Although the country passed some of the most advanced environmental protection laws in the world, they were in practice disregarded and unenforced.[19]

The First Five-Year Plan, beginning in 1928, set the stage for what followed. It relied on the stepped-up exploitation of the country's rich supply of energy, minerals, and other raw materials and on its huge pool of labor. Such rapid industrialization, however, contained within it the seeds of its own demise: eventually it became more and more difficult to obtain the necessary annual increases in labor and raw material inputs—barring revolutions in productivity, which did not occur. During the first two five-year plans (1928-1932 and 1933-1937) the economy experienced a 9.5 percent a year increase in the employed labor force; this fell to 3.8 percent in the 1960s, 2.2 percent in the 1970s, and to a mere 0.9 percent in 1980-85. Similar constraints came into play with respect to the exploitation of natural resources. For instance, according to Abel Aganbegyan, an economic adviser to Gorbachev:

> In the 1971-75 period the volume of output of the mining industry increased by 25 percent but only by 8 percent by 1981-85. This decline in growth ... was mainly connected with the worsening of the geological and economic conditions of mining. With its large-scale mining industry, currently the largest in the world, the Soviet Union is rapidly exhausting the most accessible of its natural resources.[20]

The Soviet Union was particularly wasteful in its use of material inputs. Thus with a much smaller overall industrial output than the United States, its industries nonetheless produced twice as much steel and consumed 10 percent more electricity. Worse still, despite a smaller level of agricultural production, its agriculture used about 80 percent more mineral fertilizer. Rather than spending a larger and larger proportion of its capital investment on replacing old, worn-out plant and equipment—as the advanced capitalist economies did—the Soviet Union devoted the lion's share of investment to expanding productive capacity, replacing only around 2 percent of its plant and equipment each year. The result was growing inefficiency in production. Built into this model of development, moreover, was a devotion to heavy industry as the key to rapid economic growth.[21]

By the late 1950s, it had become clear that such an unbalanced approach to economic growth was leading to dwindling growth rates as growing costs slowed down expansion. A large part of the problem was the almost exclusive attention to heavy industry at the expense of light industry and consumption. "'Production for production's sake,'" Sovietologist Moshe Lewin wrote in 1974, "certainly expressed the position of the Soviet economy, and neither the standard of living nor the national income adequately benefitted from it."[22]

Although some economic reforms were carried out in the 1960s, they were too modest to deal adequately with the growing problems of the Soviet economy, the growth rate of which declined precipitously from the 1960s on. Official data indicated an almost 50 percent drop in the growth rate between 1966-1970 and 1981-1985, from an annual average of 7.8 percent to 3.6 percent. The actual decline was much larger, due to a failure to calculate accurately the full effects of inflation. The slowdown in growth thus became what Soviet economists were to refer to as stagnation.[23]

This stagnation of the economy, together with enlarged military expenditures associated with the war in Afghanistan and an expanded arms race, exhausted the potential of the system to provide goods and services for the population. Public services deteriorated, particularly health care. The death rate per 1,000 people climbed from 7.1 in 1960 to 10.3 in 1980. Infant mortality also increased: it had reached 26 per 1,000 children under a year old in 1985, making the Soviet Union fiftieth among nations, even though it was a highly industrialized and urbanized society.[24]

During the deepening stagnation of the 1970s and early 1980s, the Soviet Union also experienced general environmental deterioration. In 1990, Alexei Yablokov, a distinguished biologist who was the most respected environmental critic in the Soviet parliament during the Gorbachev period and then became Yeltsin's top environmental adviser, cited two cases to indicate the disastrous effect that pollution was having on health:

> In the [Urals] city of Karabash (Chelyabinsk region), where harmful emissions from the smokestacks of the copper-smelting combine reach nine tons per capita per year, half the youngsters of draft age cannot be called up for army service because of the state of their health. In the Krasnodar district [of the northern Caucasus], there are rice-growing areas where the intensive use of pesticides has had such an effect on health that not a single young man could be accepted for military duty. In some farming villages of that district cancer is the only cause of death![25]

Soviet agriculture relied heavily on chemicals, often in defiance of the law. In the late 1970s and early 1980s, the tonnage of pesticides used in Soviet agriculture increased. Although banned two years earlier than in the United States, DDT was applied in certain regions with the secret permission of the ministries of agriculture and forestry. By 1987, according to Yablokov, "about 30 percent of all foodstuffs contained a concentration of pesticides dangerous to human health."

In Central Asia, cotton production that relied heavily on the intensive application of pesticides and herbicides, and on irrigation, contaminated and dried up rivers leading into the Aral Sea—once larger than Lake Huron: "As its volume shrank by two-thirds, storms carried the toxic salts from its exposed bed to fertile fields more than one thousand miles away. So much contamination by chemical wastes has been dumped into the drinking water supply that mothers in the Aral region cannot breast-feed their babies without running the risk of poisoning them."[26]

As in the United States, excessive reliance on mechanization and chemicals led to the neglect of soil conservation. Loss of precious topsoil, due to the forces of wind and water, attacked some 135 million acres of Soviet farmland between 1975 and 1990. By the late 1980s, rapid deforestation, resulting from a failure to implement sustained-yield logging practices, was an ecological menace in many parts of the eastern Soviet Union.

By the time that *glasnost* had opened up the Soviet Union sufficiently to make it possible to obtain accurate information on the environment, it had become clear that the air and water had been seriously compromised. Sulfur dioxide emissions per unit of GNP in 1988 were 2.5 times that of the United States. In extensive drinking water tests conducted in 1989, Soviet scientists discovered chemical pollutants in 19 percent of their samples and excessive bacterial levels in 12 percent. In the other former Communist regimes in Eastern Europe, ecological conditions were equally bad: Czechoslovakia and Poland had the highest levels of industrial pollution in Europe.[27]

Worst of all was the level of radioactive contamination of the environment. The April 1986 nuclear catastrophe at Chernobyl dumped more radioactive material into the atmosphere than had been released in the U.S. bombings of Hiroshima and Nagasaki. In April 1993, a report by 46 experts under the direction of Yablokov revealed that the Soviet Union had dumped 2.5 million curies of radioactive waste into the ocean (nearly 5 times the amount currently attributed to Britain's notorious Sellafield facility). "We put 17 nuclear reactors from submarines at the bottom of the sea," Yablokov reported. "Seven of them still contained fuel."[28]

There is nothing in the nature of socialism, understood as a system of production governed by the direct producers and aimed at use rather than profit, that necessitates such ecological depredation. In this respect the Soviet Union has less to tell us about the environmental traits of socialism as such than about the ecological consequences of a hierarchical, state-directed war economy caught up in a protracted Cold War. From the late 1950s on, it was apparent that the extreme centralization of decision-making, the emphasis on heavy industry in relation to light industry and consumption, and the reliance on forced drafting of resources as the main means of generating economic growth could no longer serve the system. But the needed radical reforms were not made and hence the Soviet Union entered a period of stagnation. The rise of Gorbachev in 1985 seemed at first to offer hope of a major overhaul of the system. And in the period of *glasnost* a vast new environmental movement arose that promised to address some of the worst aspects of Soviet development. The social, economic, and ecological decline of Soviet society by that time was so severe, however, that the government was no longer in a position to carry out the necessary reforms while keeping the capitalist world economy at bay. By the time of the revolutions in Eastern Europe in 1989, it was clear that capitalism had won the Cold War, and within the next few years the Soviet Union itself collapsed.

## IMPERIALISM AND ECOCIDE

As long as it persisted, the Cold War was inextricably linked to the larger question of imperialism. This was because the global conflict between capitalism and socialism was invariably at its hottest when it involved revolutions in the periphery of the capitalist world economy. Of the two major regional wars that the United States fought in Asia during the Cold War years, the most important was the Vietnam War. In its attempt to use high technology to its advantage, the United States unleashed an unprecedented level of firepower. During all of World War II, in both the European and Asian theaters, the United States had used a total of 7 million tons of munitions. By contrast, in its attempt to "pacify" Indochina, the United States used almost 15 million tons in 1964-1972 alone. Deaths from the war, when it finally ended in 1975, were well over 1 million, including over 50,000 for the United States, 180,000 for the Republic of Vietnam, and 925,000 for the North Vietnam/National Liberation Front.[29]

A central component of the U.S. strategy for winning the war was a new form of ecological warfare: defoliation. Besides destroying the forest cover, the U.S. aim was to create refugees, which would give the United States physical control over the peasantry. Herbicides were brought into the war on an experimental basis in August 1961. Agent Orange, the major defoliant used in the spraying, contained the deadly toxin dioxin. Although there was some concern within the air force that this would lead, in the words of a later air force study, to "charges of barbarism for waging a form of chemical warfare," President John F. Kennedy decided to push ahead. The treated area went from 6,000 acres in 1962 to 1.7 million acres in 1967. Over a nine-year period, 20 percent of South Vietnam's jungles and 36 percent of its mangrove forests were sprayed. Destruction of food crops was part of the same program (called Operation Ranch Hand). In 1964, 15,039 acres of crops (manioc, rice, and sweet potatoes) were sprayed; in 1967, 148,418 acres of crops were destroyed in a single year.[30]

The damage from such chemical spraying, one leading authority on herbicides in warfare wrote in 1984,

> included the death of millions of trees and often their ultimate replacement by grasses, in turn maintained to this day by subsequent periodic fires; deep, lasting inroads into the mangrove habitat; widespread site debilitation via soil erosion and loss of nutrients in solution; decimation of terrestrial wildlife primarily via destruction of their habitat; losses in freshwater fish, largely because of reduced availability of fish species; and a possible contribution to declines in the offshore fishery. The impact on the

**A refugee village near Phuc Vinh, Vietnam, in 1969, built in an area defoliated by Agent Orange. [AP/Wide World Photos]**

human population has included long-lasting neuro-intoxications, as well as the possibility of increased incidences of hepatitis, liver cancer, chromosomal damage, and the adverse outcomes of pregnancy from exposed fathers (especially spontaneous abortions and congenital malformations).[31]

In spite of the material destruction and environmental deterioration, as well as the immeasurable suffering of the people of the region, the United States was unable to win the war. In the aftermath of its defeat, there appeared what conservatives labeled the "Vietnam syndrome," whereby large parts of the U.S. population became unwilling to give their uncritical support to overseas wars. At the same time, the weakening of the U.S. balance of payments weakened the dollar (forcing President Richard Nixon to end the dollar's convertibility into gold), which marked the beginning of the end of undisputed U.S. hegemony in the world economy. This was coupled with deepening economic stagnation both in the advanced capitalist economies and the capitalist world economy as a whole. During the 1950-1973 period, the industrialized countries experienced an average annual rate of increase in Gross Domestic Product of 3.6 percent; from 1973-1989 this fell to 2.0 percent, a drop of 45 percent.[32]

In the opening provided by this weakening of U.S. power, a record number of states managed to break away from the colonial and imperial fold: Ethiopia in 1974, Portugal's African colonies (Angola, Mozambique, Guinea-Bissau) in 1974-1975, Grenada in 1979, Nicaragua in 1979, Iran in 1979, and Zimbabwe in 1980. The principal foreign policy aim of the Reagan administration, when it came into office in 1980, was to end this string of "failures," rolling back revolution across the globe and intensifying the arms race with the "evil empire" (Reagan's name for the Soviet Union). In order to accomplish this, the United States increased its officially listed "National Defense" spending (measured in constant 1982 dollars) from $1.739 trillion in the 1970s to $2.168 trillion in the 1980s, a 25 percent rise in real terms.[33] All of the nations that had carried out revolutions in the 1970s came under renewed attack, either directly by the United States or through such surrogates as South Africa, which waged a war against Angola and Mozambique.

The U.S.-financed and directed war on the Sandinista government in Nicaragua cost the U.S. treasury nearly $1 billion and resulted in an estimated $4 billion in economic and environmental damage to Nicaragua; it cost the lives of 60,000 Nicaraguans, almost 2 percent of the population. The Sandinistas had early on given their support to a revolutionary ecological program for their country. They established the country's first environmental agency—the Nicaraguan Institute for Natural Resources and the

Environment (IRENA)—which took responsibility for the nation's renewable and non-renewable resources in the nationalized sectors, such as mining, fishery, forestry, and wildlife protection. Ecology projects included soil conservation, tropical reforestation, the banning of the export of endangered species, the shift from the use of pesticides to integrated pest management, improvements in water quality, the development of appropriate technology, an extension of the national parks, and the recovery and development of indigenous seed varieties. At the same time, the Sandinistas launched parallel public health initiatives. For instance, national vaccination campaigns reduced cases of malaria by 40 percent. By 1986, Nicaragua had increased its health-care spending as a share of the national budget to 14 percent, up from 3 percent under the prerevolutionary Somoza regime. Infant mortality fell from 120 per 1,000 births in 1979 to 65 per 1,000 births in 1988, while the average life span rose from 53 to 60 years. So successful were the revolutionary ecology and public health initiatives that they were deliberately singled out for destruction by the U.S.-backed contra army. Hospitals and grain silos were attacked. More than seventy-five ecologists were killed or kidnapped. In 1983, retreating contras set fire to a reforested pine plantation in the North Atlantic Autonomous region: it burned out of control for a month, destroying more than 155 square miles of forest. In 1984, an attack destroyed the main seed storage warehouse, where indigenous seeds varieties designed to give Nicaragua agricultural self-sufficiency were being developed.[34]

The decline and fall of the Soviet Union did not bring to an end U.S. military interventions abroad. Most notably, the United States took advantage of the "new world order" brought into being by the Soviet Union's demise to intervene in Kuwait-Iraq in 1991, following the Iraqi invasion of Kuwait. Ostensibly about protecting human freedom, this war had as its major objective the control of a strategic sector of the globe and the safeguarding of oil supplies critical to the capitalist world economy. The United

States bombed 18 chemical, 10 biological, and 3 operating nuclear power plants, contaminating much of Iraq with toxic fallout. U.S. ground forces fired between 5,000 and 6,000 rounds of advanced depleted uranium (DU) armor-piercing shells. In addition, the United States and Britain launched around 50,000 DU rockets and missiles. According to a report of the United Kingdom Atomic Energy Administration, the 40 tons of radioactive debris left behind in the desert (in the form of uranium-238, which is used to make the DU weapons) has the potential of causing as many as 500,000 deaths. Since uranium-238 remains radioactive for millions of years, entire regions of Kuwait and Iraq may be indefinitely uninhabitable. According to estimates provided by Friends of the Earth International, the combined oil spills in the Gulf resulting from U.S. bombing and the alleged spilling of oil by Iraqi forces added up to as much as much as 6.8 million barrels, the largest spill in history. In addition, the burning oil wells, ignited mainly by allied aerial attacks, left long-lasting scars on the land. In fact, the oil spilled in the waters of the Persian Gulf only constituted a small part of the total oil tragedy. As science writer T.M. Hawley explains:

> Throughout the well-fire crisis, the world fixed its attention on the incredible amounts of smoke the fires pumped into the atmosphere, but most of the people who witnessed these fires firsthand say that the flames consumed only about a third to half of the oil erupting from the ground. If the Kuwait Oil Company was correct in estimating that 3 percent of the nearly 100 billion barrels of proven reserves was lost in the [well-fire] catastrophe, then at least 1.5 billion barrels either spilled directly on Kuwaiti soil or rained down [in the form on tiny droplets] on Kuwait, Saudi Arabia, Iran, Iraq and the Gulf. This is more than 130 times the largest estimate of oil spilled from tankers and export terminals into the Gulf as a result of the war and nearly 6,000 times the amount spilled in the wreck of the Exxon *Valdez.*[35]

More than ever before, the world today is divided into a hierarchy of nations. In 1976, countries classified by the World Bank as low income had an average per capita income that was

2.4 percent of that of the high-income countries. In 1982, this had dropped to 2.2 percent; in 1988, to 1.9 percent. From 1980 to 1990, the average growth of GNP per capita in the low- and middle-income countries was 52 percent of that of the advanced capitalist states.[36] Ecologically, the disparities between North and South are apparent in the different use of material inputs in these areas. In the late 1980s, the North (including the Eastern European countries) accounted for 81 percent of world paper consumption, 80 percent of its iron and steel consumption, 92 percent of cars, 81 percent of its electricity, and 70 percent of its carbon dioxide emissions. Americans consume per capita 115 times as much paper, buy 320 times as many cars, eat 52 times as much meat, and use 46 times as much electricity as their counterparts in India.[37]

In the rich countries it is common to attribute the environmental problems of the world to overpopulation. Such Malthusian views place the main onus for environmental problems not on the rich countries, where population growth is close to replacement levels, but on the poor countries, where 95 percent of the projected 3.2 billion increase in population by the year 2025 is expected to occur. Yet, as we saw in Chapter 1, the inability of most third world nations to pass through the later phase of the demographic transition and reach replacement-level fertility is a result of the structure of international inequality. "As long as famine is vivid in people's memories," feminist author Germaine Greer writes of the demographic condition in third world countries, "they will not jeopardize their chances of survival by limiting the number of children who can help to scavenge for food, children who may die. A child is never an encumbrance to a beggar: if our economic system causes the pauperization of the many, and there is no doubt that it does, it also caused the proliferation of paupers."[38] It is not overpopulation as much as the development of an economic system that places economic growth and profits before all else that has—as we shall see—brought the world to the brink of ecological disaster.

# 6

# THE VULNERABLE PLANET

In the period after 1945 the world entered a new stage of planetary crisis in which human economic activities began to affect in entirely new ways the basic conditions of life on earth. This new ecological stage was connected to the rise, earlier in the century, of monopoly capitalism, an economy dominated by large firms, and to the accompanying transformations in the relation between science and industry. Synthetic products that were not biodegradable—that could not be broken down by natural cycles—became basic elements of industrial output. Moreover, as the world economy continued to grow, the scale of human economic processes began to rival the ecological cycles of the planet, opening up as never before the possibility of planet-wide ecological disaster. Today few can doubt that the system has crossed critical thresholds of ecological sustainability, raising questions about the vulnerability of the entire planet.

## THE SCIENTIFIC-TECHNICAL REVOLUTION

During the heyday of the Industrial Revolution, the individual firm had only a small impact on the economy as a whole. The concentration and centralization (or growth and merger) of individual capitals altered this situation dramatically, changing forever the relation between firm and economy at both the national and international levels. Today over 60 percent of all U.S. manufacturing assets are owned by two hundred corporations. These changes in the economic character of the system, along with the growing internationalization of capital, constitute the essence of what has been called the monopoly stage of capitalism. Today monopoly capitalism is turning into what might be termed globalized monopoly capitalism, as a handful of multinational corporations rule over the production and finance of the entire world.[1]

Much of the concentration and centralization associated with the rise of monopoly capitalism was made possible by changes at the level of production. The most significant of these was the incorporation of scientific research and scientific management into the industrial process. "Science," according to the modern theorist of labor and technology Harry Braverman,

> is the last—and after labor the most important—social property to be turned into an adjunct of capital.... The contrast between science as a generalized social property incidental to production and science as capitalist property at the very center of production is the contrast between the Industrial Revolution, which occupied the last half of the eighteenth and the first third of the nineteenth centuries, and the scientific-technical revolution, which began in the last decades of the nineteenth century and is still going on.[2]

Although science played a large role in the early years of the Industrial Revolution, the relationship between science and industry remained indirect and diffuse. This changed in the period of the scientific-technical revolution, primarily as the result of advances in five fields: steel, coal-petroleum, chemicals, electricity, and the internal combustion engine. Germany played a leading role in these changes. As one industrial historian has noted, "It was Germany which showed the rest of the world how to make critical raw materials out of a sandbox and a pile of coal. ... IG [Farben] changed chemistry from pure research and commercial pill-rolling into a mammoth industry affecting every phase of civilization."[3]

In the United States, corporate research laboratories arose more or less in tandem with monopoly capitalism. The first research laboratory systematically organized for invention was set up by Thomas Edison in 1876. By the turn of the century, large corporations such as Eastman Kodak, B.F. Goodrich, General Electric, Bell Telephone, Westinghouse, and General Motors had each established scientific research organizations (or acquired previously independent laboratories). By 1920, there were around 300 such corporate laboratories and by 1940 some 2,200.[4]

These scientific laboratories provided a whole range of synthetic products, based on the development of new molecular arrangements, out of the essentially limitless number of those theoretically possible. This resulted in new forms of matter, many of which were created with commercial purposes in mind—from a new way of coloring fabric to a new way of killing bacteria. Unfortunately, this progress in physics and chemistry was not accompanied by an equally rapid expansion in the knowledge of how such substances might affect the environment.[5]

It is crucial to emphasize that what drove this revolution in science and technology was not simply the accumulation of scientific knowledge but the "transformation of science itself into capital."[6] This transformation was aimed at extending both the division of labor and the division of nature, and in the process both were transformed. Applied directly to the worker, this took the form of the scientific management of the labor process, the main purpose of which was to remove control over the job from the laborer and give it to management. In other words, scientific management altered labor's relation to the production process as the laborer was systematically reduced to the status of an instrument of production.

The principles of scientific management were most clearly enunciated in the early twentieth century by Frederick Winslow Taylor. Taylor was concerned with devising the theoretical tools for smashing worker resistance on the shop floor. The approach that he devised and advocated in such works as *Principles of Scientific Management*, published in 1911, centered on three principles, summarized by Braverman as: (1) the "dissociation of the labor process from the skills of the workers," (2) the "separation of conception from execution," and (3) the "use of this monopoly of knowledge to control each step of the labor process and its mode of execution."[7] In other words, complex, highly skilled labor was to be reduced to its simplest, most interchangeable—and hence cost-efficient—parts. The end result was the growing commodification of human labor and the destruction of human productive and cultural diversity.

As labor became more homogeneous, so did much of nature, which underwent a similar process of degradation. For example, as science was increasingly applied to the management of forests by profit-making businesses, the natural complexity of forest habitats was replaced by the artificial simplicity of industrial tree plantations. The goal of the scientific forester, as one authority has explained, is to simplify the forest by channeling

> the maximum amount of nutrients, water, and solar energy into the next cut of timber. He cleans up the diversity of age and size classes that are less efficient to cut, skid, process, and sell. He eliminates slow-growing and unsalable trees, underbrush, and any animals that might harm his crop. He replaces natural disorder with neat rows of carefully spaced, genetically uniform plantings of fast growing Douglas-firs. He thins and fertilizes to maximize growth. He applies herbicides and insecticides and suppresses fires to protect this crop against the ravages of nature that must be fought and defeated.

The result is a loss of biological and genetic diversity (the destruction of species and of the genetic varieties within species): industrial tree plantations are biological and genetic deserts when compared to the rich complexity of natural forests. The streams that flow through these plantations contain few fish. The variety of plants, animals, insects, and fungi is minimized. The floor of an old-growth forest is a lush carpet of vegetation; the floor of a tree plantation is almost barren by comparison. The trees themselves, which are viewed as mere commodities (i.e., so many board feet of standing timber), are "genetically improved" to allow for a lower rotation time and hence for profit maximization. Natural diversity is destroyed in the same proportion as profits are promoted.[8]

Such efficiency in the division of nature and human labor is accompanied, paradoxically, by the incorporation of useless inputs into the business process. Thus one of the primary tendencies of monopoly capitalism is "the interpenetration of the production and sales efforts," as advertising costs, superficial product changes, unnecessary packaging, and the costs of marketing in general became increasingly incorporated into the production costs of a commodity.[9]

## THE SYNTHETIC AGE

"We know that *something* went wrong in the country after World War II," Barry Commoner wrote of the United States in his bestseller *The Closing Circle* (1971), "for most of our serious pollution problems either began in the postwar years or have greatly worsened since then." That something, Commoner suggested, was "the sweeping transformation of productive technology since World War II ... productive technologies with intense impacts on the environment have displaced less destructive ones. The environmental crisis is the inevitable result of this counterecological pattern of growth." Increased throughput of energy and materials creates enormous problems associated with the depletion of resources and the treatment and disposal of wastes.

**Percent Change in the Use of Different Products in the United States, 1946-1970[10]**

|  |  |
|---|---|
| Nonreturnable soda bottles | +53,000 % |
| Synthetic fibers | + 5,980 % |
| Mercury (chlorine production) | + 3,930 % |
| Mercury (paint) | + 3,120 % |
| Air conditioner compressor units | + 2,850 % |
| Plastics | + 1,960 % |
| Fertilizer nitrogen | + 1,050 % |
| Electric housewares | + 1,040 % |
| Synthetic organic chemicals | + 950 % |
| Aluminum | + 680 % |
| Chlorine gas | + 600 % |
| Electric power | + 530 % |
| Pesticides | + 390 % |
| Wood pulp | + 313 % |
| Truck freight | + 222 % |
| Consumer electronics | + 217 % |
| Motor fuel consumption | + 190 % |
| Cement | + 150 % |
| Population | + 42 % |
| Lumber | - 1 % |
| Cotton fiber | - 7 % |
| Wool | - 42 % |

But these problems have been magnified many times over by the replacement of the products of nature with synthetics. This "technological displacement" of nature can be seen in the substitution of artificial fertilizers for organic fertilizers; the development of pesticides to replace biological forms of insect control; the use of synthetics and plastics instead of materials occurring in nature, such as cotton, wool, wood, and iron; and the replacement of soap by detergents with high phosphate content.[11]

What transpired in the post-World War II period was thus a qualitative transformation in the level of human destructiveness. Some of history's most harmful pollutants were only introduced in the 1940s and 1950s. Photochemical smog made its debut in Los Angeles in 1943. DDT first began to be used on a large scale in 1944. Nuclear fallout dates from 1945. Detergents began to displace soaps in 1946. Plastics became a major waste disposal problem only after World War II. Nuclear power and human-generated radioactive elements were a product of war industry and became industrial products in the 1950s.

During what might be termed the golden age of the postwar period (1946-1970), the physical output of food, clothing, fabrics, major household appliances, certain basic metals, and building materials (such as steel, copper, and brick) only grew at the rate of population increase, so that per capita production remained the same. For example, the per capita availability of food, whether measured in calories or protein intake, remained essentially unaltered in the U.S. over this period. Further, physical output in certain key areas actually declined over the period: cotton fibers, wool, soap, lumber, and work animal horsepower. Yet some (mainly synthetic) kinds of production increased dramatically.

While the production of basic needs—food, clothing, housing—has kept pace with the growth of population, the *types* of goods produced to meet these needs have changed dramatically. New technologies have replaced older ones. Synthetic detergents have replaced soap powder; synthetic fabrics have replaced clothing made out of natural fibers (such as cotton and wool); aluminum, plastics, and concrete have displaced steel and lumber; truck freight has displaced railroad freight; high-powered automobile engines have displaced the low-powered engines of the 1920s and 1930s; synthetic fertilizer has in effect displaced land in agricultural production; herbicides have displaced the cultivator; insecticides have displaced earlier forms of insect control.

It is therefore the *pattern* of economic growth rather than growth (or population) itself that is the chief reason for the rapid acceleration of the ecological crisis in the postwar period. In a 1972 study of the environmental impact of six pollutants (detergent phosphates, fertilizer nitrogen, nitrogen oxides, beer bottles, tetraethyl lead, and synthetic pesticides), Commoner pioneered in the introduction of the formula Environmental Impact = Population×Affluence × Technology (I = P × A × T) to assess the relative impact of the different environmental factors. He showed that P (Population) accounted for only 12 to 20 percent of the total changes in I (Impact) for these pollutants. In the case of nitrogen oxides and tetraethyl lead (both from automobile sources) around 40 percent of the changes in I were attributable to A (Affluence, defined by Commoner as economic goods/population); while in all cases other than automobile-based pollutants, A accounted for no more than 5 percent of the changes in I. T (Technology, defined as pollution/economic good) meanwhile accounted for 40 to 90 percent of all changes in I. The heightened environmental crisis of the 1970s, Commoner argued, was therefore attributable to a considerable degree to "counterecological" systems of production introduced in the postwar period.[12]

A major element in this counterecological trend has been the growth of the automobile complex. "In terms of high energy consumption, accident rates, contribution to pollution, and displacement of urban amenities," Bradford Snell stated in a famous report to a U.S. Senate committee, "motor vehicle travel is possibly the most inefficient method of transportation devised by modern man." Nevertheless, the central determinant of U.S. national transportation policy during most of the twentieth century has been a corporate strategy geared to the high profits associated with automobilization. Lavish federal funding for highways has been coupled with declining government subsidies for public transport. Moreover, at least some of the enormous present-day dependence of the United States on cars, which today account for 90 percent of all travel, can be traced to the deliberate dismantling of the nation's earlier mass transportation system. From the 1930s to the 1950s, General Motors (GM), the nation's top automobile manufacturer, operating in conjunction with Standard Oil and Firestone Tire, systematically bought up many of the nation's electric streetcar lines, converting them to buses. The number of streetcar lines dropped from 40,000 in 1936 to 5,000 in 1955. Meanwhile, GM used its monopolistic control of bus production and of the Greyhound Corporation, on the one hand, and its monopoly in the production of locomotives, on the other, to ensure the growing displacement of bus and rail traffic by private automobiles in intercity ground transport—essentially undercutting itself in intercity mass transit in order to make higher profits off increased automobile traffic. More than any other country, the United States has thus come to rely almost exclusively on cars and trucks for the ground transport of its people and goods, with disastrous consequences for the environment. *In the United States in 1988, the people/motor vehicle ratio was 1.3:1, the lowest in the world!*[13]

No less central as a counterecological force is the petrochemical industry, which creates a huge variety of synthetic products from a few starting materials, primarily petroleum and natural gas. Synthetic fibers, detergents, pesticides, and plastics are all products of the petrochemical industry, as are most toxic wastes. Today there are around 70,000 chemical preparations in use. About 400 of these have been found in the human organism, and most have never been tested for their toxic effects.[14]

The ecological impact of petrochemical production can be better understood if we take into account the way in which this industry has positioned itself in relation to both agriculture and manufacturing. On the one hand, the petrochemical industry markets agricultural chemicals. On the other hand, it produces the synthetic products that compete with farm output. Thus the giant corporations have at one and the same time molded the nation's farms into a "convenient market" and a "weakened competitor." Farming has been transformed from its ancient form, connected to ecological cycles, into a qualitatively new form of commercial enterprise known as "agribusiness."[15]

As Richard Lewontin, one of the world's leading geneticists, who teaches in the Harvard School of Public Health, and Jean-

Pierre Berlan, director of research at the French National Institute of Agronomic Research, have explained, there is an "increasing differentiation between *farming* and *agriculture*. Farming is producing wheat; agriculture is turning phosphates into bread." Although its products are essential for our survival, farming itself now accounts for only about 10 percent of the average value-added of agricultural products. Of the remaining 90 percent, 40 percent is accounted for by farm inputs (such as seeds, fertilizers, pesticides, and machinery) and 50 percent is added *after* the product leaves the farm, primarily in the form of marketing and distribution costs. The result is that a small number of large corporations, which monopolize the sale of farm inputs and the marketing and distribution of farm products, control the conditions of production in farming and reap the bulk of agricultural profits, even though farming itself is "spread over a large number of petty producers."[16]

At the heart of the Green Revolution (see Chapter 5) carried out by agribusiness in the twentieth century has been the commodification of seed production. Biotechnology has produced hybrid corn and other seed varieties that are widely touted as producing superior high-yield crops. Some scientists, however, believe that the use of the hybrid method, rather than the direct selection of high-yielding varieties from each generation and the propagation of seeds from those plants, was motivated primarily by considerations of profitability. The reason is that the use of hybrid corn seeds makes farmers purchase new seeds each year, because to pursue the traditional farming method (selecting the best plants for seeds for the following year) would in the case of hybrids result in a sharp reduction in productivity (since hybrids do not breed true and their progeny will not produce the same yields). Hybrid corn, Berlan and Lewontin write, "the flagship of the successful innovations of twentieth century agricultural research," thus "expanded the sphere of commodity production by creating a new and extraordinarily profitable commodity."[17]

Perhaps even more important, in order to cultivate these new crop varieties (which are generally less suited than earlier varieties to naturally occurring ecological conditions), large quantities of inorganic fertilizers, herbicides, and pesticides are needed, along with mechanization. The new varieties, according to ecologists Yrjö Haila and Richard Levins, are bred to perform well only if accompanied by a whole technical package. Crops such as hybrid corn therefore become the entry point for an entire model of agribusiness, one that transforms traditional farmers into economic dependents of the major agribusiness corporations.

## THE FOUR LAWS OF ECOLOGY AND ECONOMIC PRODUCTION

In order to understand the ecological impact of these trends, it is useful to look at what Barry Commoner and others have referred to as the four informal laws of ecology: (1) everything is connected to everything else, (2) everything must go somewhere, (3) nature knows best, and (4) nothing comes from nothing. The first of these informal laws, *everything is connected to everything else*, indicates how ecosystems are complex and interconnected. This complexity and interconnectedness, Haila and Levins write, "is not like that of the individual organism whose various organs have evolved and have been selected on the criterion of their contribution to the survival and fecundity of the whole." Nature is far more complex and variable and considerably more resilient than the metaphor of the evolution of an individual organism suggests. An ecosystem can lose species and undergo significant transformations without collapsing. Yet the interconnectedness of nature also means that ecological systems can experience sudden, startling catastrophes if placed under extreme stress. "The system," Commoner writes, "is stabilized by its dynamic self-compensating properties; these same properties, if overstressed, can lead to a dramatic collapse." Further, "the ecological system is an amplifier, so that a small perturbation in one place may have large, distant, long-delayed effects elsewhere."[18]

**A dump for consumer goods in Northampton, Massachusets. [Leah Melnick/Impact Visuals]**

The second law of ecology, *everything must go somewhere*, restates a basic law of thermodynamics: in nature there is no final waste, matter and energy are preserved, and the waste produced in one ecological process is recycled in another. For instance, a downed tree or log in an old-growth forest is a source of life for numerous species and an essential part of the ecosystem. Likewise, animals excrete carbon dioxide to the air and organic compounds to the soil, which help to sustain plants upon which animals will feed.

*Nature knows best*, the third informal law of ecology, Commoner writes, "holds that any major man-made change in a natural system is likely to be *detrimental to that system.*" During 5 billion years of evolution, living things developed an array of substances and reactions that together constitute the living biosphere. The modern petrochemical industry, however, suddenly created thousands of new substances that did not exist in nature. Based on the same basic patterns of carbon chemistry as natural compounds, these new substances enter readily into existing biochemical processes. But they do so in ways that are frequently destructive to life, leading to mutations, cancer, and many different forms of death and disease. "The absence of a particular substance from nature," Commoner writes, "is often a sign that it is incompatible with the chemistry of life."[19]

*Nothing comes from nothing,* the fourth informal law of ecology, expresses the fact that the exploitation of nature always carries an ecological cost. From a strict ecological standpoint, human beings are consumers more than they are producers. The second law of thermodynamics tells us that in the very process of using energy, human beings "use up" (but do not destroy) energy, in the sense that they transform it into forms that are no longer available for work. In the case of an automobile, for example, the high-grade chemical energy stored in the gasoline that fuels the car is available for useful work while the lower grade thermal energy in the automobile exhaust is not. In any transformation of energy, some of it is always degraded in this way. The ecological costs of production are therefore significant.[20]

Viewed against the backdrop offered by these four informal laws, the dominant pattern of capitalist development is clearly *counter*-ecological. Indeed, much of what characterizes capitalism as an ecohistorical system can be reduced to the following counter-ecological tendencies of the system: (1) the only lasting connection between things is the cash nexus; (2) it doesn't matter where something goes as long as it doesn't reenter the circuit of capital; (3) the self-regulating market knows best; and (4) nature's bounty is a free gift to the property owner.

The first of these counterecological tendencies, *the only lasting connection between things is the cash nexus*, expresses the fact that under capitalism all social relations between people and all the relationships of humans to nature are reduced to mere money relations. The disconnection of natural processes from each other and their extreme simplification is an inherent tendency of capitalist development. As Donald Worster explains,

> Despite many variations in time and place, the capitalistic agroecosystem shows one clear tendency over the span of modern history: a movement toward the radical simplification of the natural ecological order in the number of species found in an area and the intricacy of their interconnections.... In today's parlance we call this new kind of agroecosystem a *monoculture*, meaning a part of nature that has been reconstituted to the point that it yields a single species, which is growing on the land only because somewhere there is a strong market demand for it.[21]

The kind of reductionism characteristic of "commercial capitalism," Indian physicist and ecologist Vandana Shiva states, "is based on specialized commodity production. Uniformity in production, and the unifunctional use of natural resources, is therefore required." For example, although it is possible to use rivers ecologically and sustainably in accordance with human needs, the giant river valley projects associated with the construction of today's dams "work against, and not *with*, the logic of the river. These projects are based on reductionist assumptions [of uniformity, separability, and unifunctionality] which relate water use not to nature's processes but to the processes of revenue and profit generation."[22]

All of this is reflects the fact that cash nexus has become the sole connection between human beings and nature. With the development of the capitalist division of nature, the elements of nature are reduced to one common denominator (or bottom line): exchange value. In this respect it does not matter whether one's product is coffee, furs, petroleum, or parrot feathers, as long as there is a market.[23]

The second ecological contradiction of the system, *it doesn't matter where something goes unless it re-enters the circuit of capital*, reflects the fact that economic production under contemporary capitalist conditions is not truly a circular system (as in nature) but a linear one, running from sources to sinks—sinks that are now overflowing. The "no deposit/no return" analogue, the great ecological economist Nicholas Georgescu-Roegen has observed, "befits the businessman's view of economic life." The pollution caused by production is treated as an "externality" that is part of the costs to the firm.[24]

In precapitalist societies, much of the waste from agricultural production was recycled in close accordance with ecological laws. In a developed capitalist society, in contrast, recycling is extremely difficult because of the degree of division of nature. For instance, cattle are removed from pasture and raised in feedlots; their natural waste, rather than fertilizing the soil, becomes a serious form of pollution. Or, to take another example, plastics, which have increasingly replaced wood, steel, and other materials, are not biodegradable. In the present-day economy, Commoner writes, "goods are converted, linearly, into waste: crops into sewage; uranium into radioactive residues; petroleum and chlorine into dioxin; fossil fuels into carbon dioxide.... The end of the line is always waste, an assault on the cyclical processes that sustain the ecosphere."[25]

It is not the ecological principle that *nature knows best* but rather the counter-ecological principle that *the self-regulating market knows best* that increasingly governs all life under capitalism. For example, food is no longer viewed chiefly as a form of nutrition but as a means of earning profits, so that nutritional value is sacrificed for bulk. Intensive applications of nitrogen fertilizer unbalance the mineral composition of the soil, which in turn affects the mineral content of the vegetables grown on it. Transport and storage requirements take precedence over food quality. And in order to market agricultural produce effectively, pesticides are sometimes used simply to protect the appearance of the produce. In the end, the quality of food is debased, birds and other species are killed, and human beings are poisoned.[26]

*Nature's bounty is a free gift to the property owner*, the fourth counter-ecological tendency of capitalism, expresses the fact that the ecological costs associated with the appropriation of natural resources and energy are rarely factored into the economic equation. Classical liberal economics, Marx argued, saw nature as a "gratuitous" gain for capital. Nowhere in establishment economic models does one find an adequate accounting of nature's contribution. "Capitalism," as the great environmental economist K. William Kapp contended, "must be regarded as an economy of unpaid costs, 'unpaid' in so far as a substantial portion of the actual costs of production remain unaccounted for in entrepreneurial outlays; instead they are shifted to, and ultimately borne by, third persons or by the community as a whole." For example, the air pollution caused by a factory is not treated as a cost of production internal to that factory. Rather it is viewed as an external cost to be borne by nature and society.[27]

By failing to place any real value on natural wealth, capitalism maximizes the throughput of raw materials and energy because the greater this flow—from extraction through the delivery of the final product to the consumer—the greater the chance of generating profits. And by selectively focusing on minimizing labor inputs, the system promotes energy-using and capital-intensive high technologies. All of this translates into faster depletion of nonrenewable resources and more wastes dumped into the environment. For instance, since World War II, plastics have increasingly displaced leather in the production of such items as purses and shoes. To produce the same value of output, the plastics industry uses only about a quarter of the amount of labor used by leather manufacture, but it uses ten times as much capital and thirty times as much energy. The substitution of plastics for leather in the production of these items has therefore meant less demand for labor, more demand for capital and energy, and greater environmental pollution.[28]

The foregoing contradictions between ecology and the economy can all be reduced to the fact that the profit-making relation has become to a startling degree the sole connection between human beings and between human beings and nature. This means that while we can envision more sustainable forms of technology that would solve much of the environmental problem, the development and implementation of these technologies is blocked by the mode of production—by capitalism and capitalists. Large corporations make the major decisions about the technology we use, and the sole lens that they consider in arriving at their decisions is profitability. In explaining why Detroit automakers prefer to make large, gas-guzzling cars, Henry Ford II stated simply "minicars make miniprofits." The same point was made more explicitly by John Z. DeLorean, a former General Motors executive, who stated, "When we should have been planning switches to smaller, more fuel-efficient, lighter cars in the late 1960s in response to growing demand in the marketplace, GM management refused because 'we make more money on big cars.'"[29]

Underlying the general counter-ecological approach to production depicted here is the question of growth. An exponential growth dynamic is inherent in capitalism, a system whereby money is exchanged for commodities, which are then exchanged for more money on an ever increasing scale. "As economists from Adam Smith and Marx through Keynes have pointed out," Robert Heilbroner has observed, "a 'stationary' capitalism is subject to a falling rate of profit as the investment opportunities of the system are used up. Hence, in the absence of an expansionary frontier, the investment drive slows down and a deflationary spiral of incomes and employment begins." What this means is that capitalism cannot exist without constantly expanding the scale of production: any interruption in this process will take the form of an economic crisis. Yet in the late twentieth century there is every reason to believe that the kind of rapid economic growth that the system has demanded in order to sustain its very existence is no longer ecologically sustainable.[30]

# 7

# THE SOCIALIZATION OF NATURE

## THE FAILURE OF ECOLOGICAL REFORM

The history of ecological struggle over the last thirty years presents us with a clear picture of what can and cannot be expected of ecological reform within the confines of the system. The publication in 1962 of Rachel Carson's *Silent Spring*, which raised the alarm about the poisonous effects of pesticides, represented a turning point in the U.S. environmental movement. Soon large sections of the population were waking up to a host of ecological dangers symbolized by DDT, L.A. smog, toxic wastes in Love Canal, the death of the Great Lakes, acid rain, the energy crisis, oil spills, the Three Mile Island nuclear meltdown, and the clearcutting of the national forests. The public outcry gained national prominence with the first Earth Day in April

1970, and a torrent of new federal laws were passed to "regulate" the environment.

The environment, however, continued to decline. Air pollution emissions per annum improved only modestly (with the exception of lead) between 1970 and 1982, with no improvement after that. In 1970, there were 98 million cars and trucks emitting pollution in the United States; by 1988, 170 million. Over the same period, the number of motor vehicle miles traveled rose by 72 percent, to 1.9 trillion. Water quality in most rivers has failed to improve, while concentrations of some pollutants—nitrate, arsenic, and cadmium—have increased sharply. The number of oil spills in and around U.S. waters increased by 196 percent between 1970 and 1985, and their total volume by 57 percent. The Environmental Protection Agency has reported that the rate of oxygen depletion in the central basin of Lake Erie went up by 15 percent between 1970 and 1980.[1]

Despite increased regulation, the overall pesticide problem is worse than it was when Rachel Carson published *Silent Spring*. In the words of Shirley Briggs, executive director of the Rachel Carson Council:

> *Silent Spring* recorded a rise in the production of pesticide active ingredients in the United States from 124,259,000 pounds in 1947 to 637,666,000 pounds in 1960. By 1986, according to Environmental Protection Agency (EPA) figures, production had risen to 1.5 billion ... pounds for the range of products cited by Rachel Carson, and U.S. use was about 1.09 billion pounds.... If wood preservatives (fungicides), disinfectants and sulphur are taken into account, the figure for U.S. pesticide usage in 1988 is 2.7 billion pounds. U.S. pesticide production (not including the latter three categories) is about one-quarter of the world total, so the annual burden on the earth must be about 6 billion pounds of these products.[2]

Further, the accumulation of toxic chemicals in the environment is also increasing rapidly. U.S. industry reported dumping 20 billion pounds of toxic chemicals into the air, water, and land in 1987. Yet the real figure, according to the Congressional Office

**In the United States in the 1940s and 1950s, DDT was sprayed directly on sheep and crops. [AP/Wide World Photos]**

of Technology Assessment, may be closer to 400 billion pounds. The government has confirmed some 30,000 hazardous waste sites throughout the country—a list that is growing rapidly. Over 560 million tons of hazardous waste are generated by U.S. industry each year—2 tons for each member of the population. There are over 200 industrial chemicals and pesticides "commonly found in the body tissue of 95 percent of Americans tested."[3]

Even the Endangered Species Act, the strongest wildlife protection act in the world, has failed so dramatically in the context of the rapid destruction of forests, wetlands, rivers, coastal estuaries, and other wildlife habitats, on the one hand, and the slow speed with which species are officially listed for protection, on the other, that some environmentalists have labeled it a "chronicle of extinction."[4]

This story of environmental decline is repeated at the global level. According to the Worldwatch Institute, "The health of the planet has deteriorated dangerously during the twenty years" since the U.N. meeting at Stockholm in 1972 that officially launched the global environmental movement. "Modern humanity," ecologist Edward Goldsmith has written,

> is rapidly destroying the natural world on which it depends for survival. Everywhere on our planet, the picture is the same. Forests are being cut down, wetlands drained, coral reefs grubbed up, agricultural lands eroded, salinized, desertified, or simply paved over. Pollution is now generalized—our groundwater, streams, rivers, estuaries, seas and oceans, the air we breathe, the food we eat, are all affected. Just about every living creature on earth now contains in its body traces of agricultural and industrial chemicals—many of which are known or suspected carcinogens or mutagens.[5]

Despite the growth of environmentalism in Britain, the U.K.'s nuclear complex at Sellafield (on the Irish Sea in the north) continues to be both the world's foremost commercial producer of plutonium and one of the world's leading sources of radioactive contamination. Sellafield has had around 300 accidents, including a core fire in 1957 that was the worst accident to occur in a nuclear reactor prior to Chernobyl. It is the center of an international trade in nuclear wastes, processing radioactive wastes from West Germany, Japan, and other countries, which in return purchase British nuclear technology and plutonium. For decades Sellafield has been flushing radioactive residues down a mile-and-half-long pipeline into the Irish Sea, "creating an underwater 'lake' of wastes, including, according to the British government, one-quarter ton of plutonium, which returns to shore in windborne spray and spume, and in the tides, and in fish and seaweed and flotsam, and which concentrates in inlets and estuaries."[6]

This failure to prevent the increased destruction of the biosphere can be traced mainly to the logic of profit-oriented economic expansion in a finite world. Where legislation has been adopted to protect the environment from the worst forms of pollution, it has usually taken the form of attempts to regulate harmful emissions while leaving the structure of production and profit-making intact. Hence, only rarely—such as in the banning of DDT or the removal of lead additives from fuel—is the problem dealt with at its source, through prevention, although when this happens the result is dramatic environmental improvement. More often, an attempt is made to "control" the problem and to reduce the risk to human beings and nature to "acceptable levels." For instance, control devices are installed to trap or destroy pollutants before they enter the atmosphere—the catalytic convertor in an automobile destroys carbon monoxide and unused gasoline, and power plants have scrubbers installed that trap sulfur dioxide and gas. Such devices, however, are costly and never fully effective. Even with a catalytic convertor for every car, more cars still mean more pollution. The number of passenger cars on roads around the world has more than doubled between 1970 and 1990, from around 200 million to some 450 million vehicles.[7]

In the Reagan years, the U.S. government deliberately sought to weaken environmental regulation as part of an attempt to restructure the capitalist economy in the face of economic stagnation. For instance, no sooner did Reagan come into office than he proposed doubling the rate of cutting in the old-growth forests of the Pacific Northwest. He also eliminated the entire staff of the Council of Environmental Quality, while the Environmental Protection Agency's workforce was cut by a quarter, its operating budget by a third, and its research funding by half. All of this made it possible for corporations to step up the pace of environmental destruction.[8]

## ENVIRONMENTAL REVOLUTION

Given these environmental failures of the last thirty years, it is obvious that a more radical response is needed. Humanity is clearly approaching a turning point in its long history, one marked by the stark choice between environmental revolution and environmental decline. According to Worldwatch's *State of the World 1992:*

> Building a sustainable future depends on restructuring the global economy, major shifts in human reproductive behavior, and dramatic changes in values and life-styles. Doing all this quickly adds up to a revolution, one defined by the need to restore and preserve the earth's environmental systems. If this Environmental Revolution succeeds, it will rank with the Agricultural and Industrial Revolutions as one of the great economic and social transformations in history.[9]

But what is to be the nature of this "environmental revolution"? For the Worldwatch Institute, and for mainstream environmentalism in general, it involves far-reaching changes in "population, technology, and lifestyle," and a general restructuring of the world economy in order to promote what is called "sustainable development." Government will have to play a more active role in environmental regulation, corporations will have to reform to become more environmentally responsible, and a "green" industrial strategy will have to be devised to ensure that development remains sustainable.[10]

What is conspicuously missing from this conception of an environmental revolution, however, is any recognition of the fact that in order to halt, or even significantly slow down, the rate of environmental deterioration, capitalist commodity society will have to give way to environmental necessity. In other words, an effective environmental movement, capable of addressing the rapid destruction of the ecology of the planet, can only develop if root problems of production, distribution, technology, and growth are dealt with on a global scale. Such hesitations before the issue of the social governance of production reflect a long-established pattern within the dominant environmental paradigm. Characteristic of mainstream environmentalism since its inception, as Hans-Magnus Enzensberger has observed, is a form of thought that smacks of a preacher's sermon, in which "the horror of the predicted catastrophe contrasts sharply with the mildness of the admonition with which we are allowed to escape."[11]

Nowadays the main such admonition is that we have failed to develop "sustainably." "Sustainable development" is thus offered as the magic solution to the world's catastrophic ecological problems. Yet this resolution is shaped in such a way as to make environmental progress synonymous with the sustainable development of *capitalism.*

The most prestigious establishment presentation of this perspective remains the 1987 Brundtland Commission report, *Our Common Future,* produced by the World Commission on Environment and Development. "Sustainable development," the Brundtland report states, means "development that meets the needs of the present without compromising the ability of future generations to meet their own needs." "The Commission's overall assessment," the report continues, "is that the international economy must speed up world growth while respecting environmental restraints." The Commission thus called for "more rapid economic growth in both industrial and developing countries, freer market access for products of developing countries, lower interest rates, greater technology transfer, and significantly larger capital flows, both concessional and commercial." Growth (particularly in less-developed countries) by a factor of five to ten is necessary, it is argued, in order to bring those countries up to the level of consumption of the industrialized countries by the time that population levels are expected to level off in the middle of the next century. Meanwhile, the rate of growth of the industrialized countries must speed up as well. All of this is to be accomplished while promoting "less material-intensive and energy-intensive technology," creating more globally equitable development, and reducing world population pressures.[12]

The Brundtland Commission's insistence on the need for faster growth, larger capital flows, and increased access to the natural resources of underdeveloped countries all reflect a commitment to the needs of capital rather than the environment.

Logically, in order to be physically sustainable, an ecohistorical formation has to meet three conditions: (1) the rate of utilization of *renewable* resources has to be kept down to the rate of their regeneration; (2) the rate of utilization of *nonrenewable* resources cannot exceed the rate at which alternative sustainable resources are developed; and (3) pollution and habitat destruction cannot exceed the "assimilative capacity of the environment." Yet to achieve these ends, according to current ecological knowledge, we must not simply slow down present economic growth trends but *reverse* them. Nothing in the history of capitalism suggests that this will happen.[13]

The Brundtland report bears witness to the enormous difficulties inherent in creating a sustainable development strategy in the context of capitalist society. When the report was released, the world was faced by two overlapping crises: world economic stagnation and deepening global environmental crisis. The report itself referred to "economic stagnation" in the underdeveloped countries, while average annual per capita growth rates in the advanced capitalist world had fallen 45 percent between 1950-1973 and 1973-1989. In this situation, it was economic stagnation and the necessity of restructuring the world economy in order to open the way for ever greater market dominance over the human and natural conditions of production that took precedence. In the end, therefore, the only "solutions" to the ecological dilemma offered by Brundtland were greater innovation, population control, and faster growth. In like fashion, the North American Free Trade Agreement (between Canada, Mexico, and the United States) was offered by many commentators as the solution to Mexico's long-standing problems of underdevelopment *and* environmental decay—even though the primary purpose of the agreement was to promote accumulation, not ecological sustainability.[14]

It is true that the impoverishment and misery of billions of people cannot be surmounted without major advances in agricultural and industrial output, increased productivity, and scientific and technical progress. These necessities, however, must be disconnected from the demand for accumulation for accumulation's sake or (as in the Soviet Union) production for production's sake, which have too often guided economic development. One of the basic contradictions of the capitalist economy, according to radical philosopher István Mészáros, "is that it cannot separate 'advance' from *destruction*, nor 'progress' from *waste*—however catastrophic the results." If priority were given to meeting the needs of the poorest people in the most underdeveloped regions and to protecting the environment, development would necessarily take on a radically different character. But to accomplish this shift in priorities it is necessary to break with the model provided by the consumption standards and way of life of the advanced capitalist states.[15]

How much can be accomplished even in the context of reduced growth rates and low incomes can be seen by looking at states that have emphasized redistributive policies, such as Kerala (a state within India), Sri Lanka, China, and Cuba. Kerala, with a population of 29 million (greater than that of Canada) has a per capita income that is only 60 percent that of India as a whole. Nevertheless, due to a history of mass struggle, which brought a communist party (or parties) to power for long periods, it has made startling progress in areas of land reform, distribution, nutrition, health, and education. Infant mortality in Kerala is 27 per thousand, compared with 86 per thousand in India as a whole and 106 per thousand in countries at the same per capita income level as Kerala. Life expectancy in India is 57 years, while in Kerala it is now more than 70 years—far exceeding that of much wealthier countries like Saudi Arabia and Brazil and approaching that of advanced capitalist states like the United States. "Such an accomplishment in the face of very low income," according to Harvard economist Amartya Sen, writing in *Scientific American*, "is the result of the expansion of public education, social epidemiological care, personal medical services and subsidized nutrition." Similarly, in Sri Lanka, China, and Cuba, programs of public education, socially guaranteed medical services, and state subsidized food supplies have contributed to high life expectancies despite low per capita incomes.[16]

## TWO LOGICS

As the authors of *Europe's Green Alternative* have written, there are two "opposing logics" at work in the world today: "On the one side, economics divorced from all other considerations; and on the other, life and society." Where the economic logic of capitalism is concerned, there is no doubt that the dominant tendencies remain what they have always been: the accumulation of wealth at one pole and the accumulation of relative misery and degradation at the other. In this process "ecosystems are inert worlds to be pillaged and poisoned at will. Human beings are defined by their capacity to work and consume, transformed from free subjects into objects."[17]

Fortunately for the world, however, there are other forces at work, even within the context of capitalist society, that limit the system's capacity to impose its economic logic on nature and humanity. Market society, as the great economic anthropologist Karl Polanyi observed, was characterized from the beginning by a "double movement" that

> can be personified as the action of two organizing principles in society, each of them setting itself specific institutional aims, having the support of definite social forces and using its own distinctive methods. The one was the principle of *economic liberalism*, aiming at the establishment of a self-regulating market, relying on the support of the trading classes and using largely

> *laissez-faire* and free trade as its methods; the other was the principle of *social protection* aiming at the conservation of man and nature as well as productive organization, relying on the varying support of those most immediately affected by the deleterious action of the market—primarily, but not exclusively the working and landed classes—and using protective legislation, restrictive associations, and other instruments of intervention as its methods.[18]

The second part of this "double movement," the "principle of social protection," is for capital an externally imposed necessity—one that stems from social rather than market forces. Thus the new popular movement aimed at stopping the uncontrolled disposal of hazardous wastes in the United States, as Lois Gibbs, one of its leaders, has stated, does not emanate from the vested interests, but is made up of

> tens, if not hundreds of thousands of people—blue collar workers, housewives, people of color, farmers, small business owners, homeowners, renters, urban people, and rural people. They're young and old and every age in between. They have mainly low and moderate incomes.... The thing they have in common is a fear and concern about how this country manages hazardous waste.... They've found out that they're living near one of the over 100,000 hazardous waste sites that exist across the country. Or they've found that their community has been chosen to receive a share of the 250 million tons of hazardous waste the United States produces every year.... So the time comes when concerned community people decide they have to fight. Parents fight to protect their children. Working people fight to protect all of the things they have worked so hard to get. Homeowners fight because they feel they're being backed into a corner. Farmers and business owners fight to protect their lives and health. And over it all is a fight for justice.[19]

The increasingly radical nature of the grassroots environmental movement in recent years derives from the wider protections needed by society and nature in the face of the growing commodification of life. Traditionally, the mainstream U.S. environmental groups, known as the Group of Ten (and including organizations like the Sierra Club, the Audubon Society, the National Wildlife Federation, the Natural Resources Defense Council, and the Wilderness Society), have emphasized lobbying action to promote new legislation in the area of conservation. Most of these organizations have drawn their memberships and staffs predominantly from the white, educated middle strata, and they have relied heavily on corporate funding. For example, members and subscribers only accounted for 37 percent of the total funding of the National Wildlife Federation in 1988. The rest came from corporations, primarily Amoco, ARCO, CocaCola, Dow, Duke Power, Dupont, Exxon, GE, GM, IBM, Mobil, Monsanto, Tenneco, USX, Waste Management, Westinghouse, and Weyerhaeuser. Indicative of the general attitude of such groups is the fact that in their *Blueprint for the Environment,* presented to George Bush when he became president, the established environmental groups presented 750 detailed recommendations aimed at every Cabinet department but two—the Department of Housing and Urban Development and the Department of Labor—presumably because these were seen as being geared to issues of race and class.[20]

In sharp contrast, the newly emerging ecology movement of the 1980s—which arose in response both to the worsening environmental crisis and the failures of mainstream environmentalism—has been associated with direct action on behalf of the environment and grassroots mobilization to protect communities. Organizations like Greenpeace, the Earth Island Institute, and Earth First! have increasingly distinguished themselves from the older mainstream groups by taking more radical stands on a host of environmental issues. In 1983, four Earth First!ers appeared out of nowhere in the Siskiyou National Forest in Oregon and took their stand between a running bulldozer and a tree. Before long environmentalists were sitting on company dynamite to prevent blasting, tree sitting, chaining themselves to timber equipment, and forming human barricades on logging roads by setting their feet in cement-filled ditches or inserting themselves in rock piles.

But wilderness direct-action groups like Earth First! are rooted in romantic notions of nature and of a return to a simpler life.[21] The emerging urban/community-based environmental coalitions, in contrast, tend to be broad alliances founded on the self-mobilization of peoples of color, women, and workers. Among the most notable in this respect are the Southwest Organizing Project/Southwest Network for Environmental and Economic Justice, based in Albuquerque, New Mexico; the Gulf Coast Tenants Leadership Development Project; the Citizen's Clearinghouse for Hazardous Wastes and the National Toxics Campaign; the L.A. Labor/Community WATCHDOG Group; and the White Earth Recovery Project in White Earth, Minnesota. These and similar organizations form the backbone of what is commonly referred to as the "environmental justice movement." Its principal concerns have been the relationship between environmental degradation and social and economic injustice, particularly in relation to race, gender, and class oppression.

This movement is closely allied with certain unions that have historically fought for improvements in environmental conditions (particularly in the area of occupational health). The most important of these has been the Oil Chemical and Atomic Workers Union (OCAW), which has recently advocated a "Superfund for Workers" that would help pay for a change of career for those workers who lose their jobs through environmental or occupational regulations or the decline of hazardous military-based production. When the OCAW was faced in 1984 with a lockout in Geismar, Louisiana, by the giant West German chemical company, BASF, the union countered with an environmental campaign against BASF, in which OCAW organizers helped to establish such organizations as the largely African American Ascension Parish Residents Against Toxic Pollution, the Louisiana Workers Against Toxic Chemical Hazards, the Geismar-based Clean Air and Water Group, and the Louisiana Coalition for Tax Justice (which challenged tax breaks for companies producing toxic wastes).[22]

The struggle against environmental racism—defined as the institutionalization of environmental hazards in ways that disproportionately impact people of color—has been an important part of the environmental justice movement. In 1982, in Warren County, North Carolina (an area 60 percent African American and 4 percent American Indian), people rose up in protest against the placement of a proposed PCB (polychlorinated biphenyls) disposal site in their community. In this struggle, 500 predominantly African American protestors, including Benjamin Chavis, executive director of the United Church of Christ Commission for Racial Justice (now of the NAACP), were arrested. Although ultimately unsuccessful, this protest helped to draw regional and national attention to the phenomenon of environmental racism.

In Los Angeles, over 70 percent of African Americans, 50 percent of Latinos, but only 34 percent of whites live in the areas with the greatest air pollution. A 1987 United Church of Christ report entitled *Toxic Wastes and Race in the United States* concluded that communities with "greater minority percentages of the population are more likely to be sites of commercial hazardous waste facilities." Organizing against this environmental racism has therefore become a major factor in the growth of the new urban-based environmental coalitions.[23]

These struggles, because of their inherently radical nature, ultimately lead back to structures of production and property—to the basic elements of the system itself. "Oppressed people," Scott Douglas, director of the Greater Birmingham Ministries explains, "do not have compartmentalized problems." They don't make hard and fast distinctions between such environmental issues as the location of a new incinerator in their communities or lead poisoning in their schools, on the one hand, and social questions of poverty, underfunding of schools, lack of jobs, and absence of day-care, on the other. The movement in the Southwest, as Richard Moore, co-chair of the Southwest Network for

**Protestors demonstrating against a new incinerator in Los Angeles. [Crossroads Magazine]**

Environmental and Economic Justice has pointed out, has tried to link environmental issues with the needs of workers: "We've closed down plants.... a plant may have to go: it's killed people inside, and has also poisoned our groundwater and our air and our children on the outside. But we went through a process first, attempting to bring workers into the decision."[24]

This refusal to separate the struggle for the environment from issues of social justice has made the environmental justice movement sharply critical of the history and logic of U.S. capitalism. In a famous protest letter written to the Group of Ten environmental organizations in 1990, the Southwest Organizing Project declared:

> For centuries people of color in our region have been subjected to racist and genocidal practices including the theft of lands and water, the murder of innocent people, and the degradation of our environment. Mining companies extract minerals leaving economically depressed communities and poisoned soil and water. The U.S. military takes lands for weapons production, testing and storage, contaminating surrounding communities and placing minority workers in the most highly radioactive and toxic worksites. Industrial and municipal dumps are intentionally placed in communities of color, disrupting our cultural lifestyle and threatening our communities' futures. Workers in the fields are dying and babies are born disfigured as a result of pesticide spraying.[25]

The letter went on to attack the systematic "expropriation of Third World Resources" and to demand an internationalist strategy that will "create a global environmental movement that protects us all."

Such internationalist alliances seem increasingly feasible. In recent decades powerful environmental justice movements in the third world have challenged the pattern of capitalist development. India's Chipko (or tree-hugging) movement, located in the sub-Himalayan region, is a form of struggle led by women that has its roots in ancient Indian culture. In the 1970s, women fought to eliminate cash cropping and to restore traditional ecological uses of the forest for food (providing fruits, roots, tubers, seeds, leaves, petals, and sepals), fuel, fodder, water, fertilizer, and medicine. In the process of embracing trees for these purposes, they came into conflict with business and government.[26] In the words of Indian physicist and ecologist Vandana Shiva, "There are in India today two paradigms of forestry—one life-enhancing, the other life-destroying. The life-enhancing paradigm emerges from the forest and the feminine principle; the life-destroying one from the factory and the market.... Since the maximizing of profits is consequent upon the destruction of conditions of renewability, the two paradigms are cognitively and ecologically incommensurable."[27]

In the face of the rapid destruction of the tropical rainforest in Brazil, an Amazonian rubber tapper named Chico Mendes organized his fellow rubber tappers against the landowning interests who were destroying the forest. In this fierce struggle, the rubber tappers allied themselves with indigenous people in defense of the forest. By 1989, the year after Mendes was murdered by local ranchers, the rubber tappers estimated that they had preserved some 3 million acres of rubber trees.[28]

More than simply struggling for extractive reserves that would allow both the ecology and their culture to persist, the rubber tappers were struggling for fundamentally different property relations. "Chico Mendes's career as an environmentalist," Susanna Hecht and Alexander Cockburn wrote in *The Fate of the Forest*,

> cannot be divorced from his active life as an extremely radical political militant.... To grasp the concrete implications of a phrase like "socialist ecology" consider the implications of the manifestoes promulgated by the rubber tappers and the Forest People's Alliance.... They emphasize that plans for the development of the region should be based on the culture and traditions of forest peoples. They take as axiomatic the preservation of the environment and the improvement of the quality of life.... Forest people seek legal recognition of native lands and extractive reserves held under the principle of collective property, worked as individual holdings with individual returns.... They oppose a political economy that favors large owners who impose social and ecological ruin on the region. Once seen through the eyes of local people, the schemes of development agencies and functionaries based in Brasilia or in Washington take on a very different aspect. At the level of everyday life forest people demand control of the relations of production and of the distribution of the fruits of their labor.[29]

This struggle of Brazilian forest peoples to create extractive reserves and to recognize legally native lands that will be held on collective principles serves to illustrate the historical process that some environmentalists have called the "socialization of nature." Everywhere in the world, in response to growing environmental depredations and the accompanying destruction of cultures, people are beginning to reassert their traditional right as human beings to share in the common benefits of nature. Moreover, the protection of plants, animals, landscapes, and large ecosystems requires "centralized management ... on a nationwide scale." As a result, struggles are taking place on many different levels, all aimed at removing nature from the realm of free exploitation and putting it under public protection.[30]

The relentless privatization of nature and production, by making all turning back impossible and all continued forward movement within the system equally impossible, leaves little option—if human beings are to continue to advance—other than the socialization of nature and production. Only in this way can the conditions of life and human existence be safeguarded. Since work constitutes the basis of the human relation to nature, the socialization of nature can only be fully realized if accompanied by the socialization of production. Environmental revolution thus necessitates social revolution. Only through the democratically organized social governance of both production and nature on a global scale is there any meaningful hope—and even then no guarantee—that the world will be cared for in common and in the interest of generations still to come, rather than simply exploited for individual short-term gain.

# AFTERWORD

*The Vulnerable Planet* was published only six years ago, and it might be said that a new edition is hardly warranted. The main data and trends are still pertinent, despite such changes as a growth in world population to 6 billion from the 5.5 billion of the early 1990s. The central argument seems more relevant than ever. The reception to the book has more than fulfilled my own wish as an author that it would play a role in inspiring individuals to join the struggle for a sustainable and just society. It is with some trepidation, then, that I approached the issue of writing an afterword to a new edition.

Yet at this millennial juncture in human history, I am deeply conscious of the need to throw some additional light on the complex issue of time scale that lies at the very heart of the environmental crisis and the subject matter of *The Vulnerable Planet.* Although a date like the year 2000 is many ways an arbitrary convention, resulting from the widespread adoption of the calendar introduced by Julius Caesar in 45 B.C., it encourages us to place our lives in the context of centuries and epochs rather than mere years and decades. It therefore raises, in a very poignant way, the question of the intersection of our biographies with historical time.

In writing this book I deliberately employed three different time scales. The initial perspective was that of the millennia that describe the period of written history, stretching from the rise of the earliest tributary civilizations to the advent of capitalism more than five thousand years later. This was subsequently narrowed down to a time scale of a few centuries when capitalism entered the picture some five hundred years ago, followed by machine capitalism two hundred years ago. Later the time frame was narrowed down even further, to a scale of decades, when the question of "the synthetic age" arose around fifty years ago. Such an approach allowed me to devote increasing space to historical developments as we moved closer to the present.

But much more was involved here than mere convenience of exposition. More important was a conviction that the relevant time frame for viewing environmental problems was rapidly contracting as we moved closer to the present. This was strongly impressed upon my mind by a vivid sense of the exponential growth tendencies of capitalist society, with its dynamic of accumulation and its tendency to revolutionize all fixed, fast-frozen relations. We can get a clear sense of the enormous impact of this stepped-up pace of change if we recognize that over the course of geological time there have been five great periods of mass extinction in which 65 percent or more of all species have died out (most recently the end-Cretaceous extinction of the dinosaurs). Today some leading scientists are pointing to the possibility of a "sixth extinction"—this time at the hand of humanity. Half of the species on earth are now threatened with extinction on a scale of *decades* as a result of the rapid elimination of the world's tropical rainforests alone.[1]

Here it is useful to draw on an insight that originated in the nineteenth century. Only two years after Karl Marx completed the first volume of *Capital* and five years after George Perkins Marsh published *Man and Nature*, the great conservative historian Jacob Burckhardt introduced the concept of the "acceleration of historical processes" in order to define the phenomenon of "historical crisis." For Burckhardt a "historical crisis" exists when "a crisis in the whole state of things is produced, involving whole epochs and all or many peoples of the same civilization. . . . The historical process is suddenly accelerated in terrifying fashion. Developments which otherwise take centuries seem to flit by like phantoms in months or weeks, and are fulfilled."[2] Burckhardt was of course aware of the enormous changes taking place in the area of production (including technological changes) which in many ways set the stage for this "acceleration of historical processes," but his chief concern was the revolution in social relations—particularly with regard to the state, religion, and culture.

As a romantic pessimist Burckhardt could identify neither with the kind of working-class revolt evident in his time nor with the purely money-grubbing society growing up around him, so he sought solace in the study of cultural history. Still, his concepts of the "acceleration of history" and "historical crisis" had certain features that were parallel to the thinking of his great contemporaries Marx and Marsh. For Marx, human productive forces and the human relation to nature were changing at a bewildering rate in the capitalist epoch. But as a revolutionary optimist (quite the opposite in this respect to Burckhardt) he identified with the social revolutions and class struggles occurring in his time and saw this as pointing beyond the mere triumph of bourgeois society. Marx believed that it would be possible, not all at once but through a process of historical struggle, to create a society aimed at the transcendence of human self-alienation and of the alienation of human beings from nature. Marsh, in contrast to both Burckhardt and Marx, did not present a social vision. He described only the tremendous pace and growing spatial scale with which the new industrial society was transforming and degrading the earth. The end result of such ecological destruction, he warned, might be the destruction of human society itself.

Recently the Worldwatch Institute, the leading think tank on global environmental change, has reintroduced the concept of "the acceleration of history" (though without any mention of Burckhardt, Marx or Marsh) in order to explain the phenomenon of environmental crisis as it presents itself in our time. According to the Worldwatch analysts, the acceleration of human history began some forty thousand years ago when ever-more sophisticated tools were developed for hunting, cooking, and other tasks related to subsistence production. Another burst of accelerated change occurred ten thousand years ago with the rise of settled agriculture. A third acceleration occurred in the middle centuries of the current millennium in the form of a scientific and technological revolution, eventually leading to the Industrial Revolution of the eighteenth century. Finally, over the last century a technological and scientific explosion has led to an "accelerating pace of change" revolutionizing "virtually every field of human activity"—manifested in what the Worldwatch analysts call "the growth century."[3]

These analysts end with the observation that "history will undoubtedly continue to accelerate" but "historical trends will have to move in a new direction" if we are not to destroy the ecological bases of human development. At present certain critical ecological thresholds or limits are being fast approached (in some cases crossed) that portend global catastrophe if we do not mend our ways. "Are we headed," they ask, "for a world in which accelerating change outstrips our management capacity, overwhelms our political institutions, and leads to extensive breakdown of the ecological systems on which the economy depends? . . . Even in a high-tech information age, human societies cannot continue to prosper while the natural world is progressively degraded."

But although the Worldwatch analysts raise the question of "the acceleration of history" in this dramatic way, the solutions they offer consist entirely of technological and demographic adjustments, as if this will be sufficient. Calling for a "new economy," they tell us that this simply means moving from fossil fuels to solar power, from cars to bicycles and railroads, from a throwaway society to one dedicated to recycling. The overall approach is one of improved management engineered by technological elites within the present global socio-economic order.

Not even economic growth is seriously addressed. Rather we are told that enlightened leadership within multinational corporations may offer the way out: a new "eco-industrial world." "Corporate behemoths, such as BP, General Motors, and Dupont, that rose on the crest of the fossil fuel revolution could capture many of the new opportunities" represented by a switch to a sustainable society. "Or they could be elbowed aside by the Microsofts of the new technological generation." Bill Ford, who became Chairman of the Ford Motor Company in 1998, we are told, is a self-described "passionate environmentalist" who is predicting the eventual demise of the automobile. Such changes in thinking at the top, coupled with changes in consumer behavior encouraged by green taxes, will be adequate, we are led to believe, to cope with the acceleration of history and the terrifying environmental threat that it has engendered. The fact that corporations are essentially enormous engines of concentrated economic expansion, and that CEOs are obligated to serve these interests, is passed by in silence.[4]

Indeed, what is most notable about this analysis is what it leaves out. In this discussion of the acceleration of history there is no mention of historical crisis in the Burckhardtian sense, of "a crisis in the whole state of things . . . involving whole epochs." Instead, historical crisis in this sense is seemingly a phenomenon of the past, not deserving mention, except as part of a general threat of apocalypse encompassing all cultures and species. All that is deemed necessary is to discuss the technological acceleration of history, which is subject to purely utilitarian, technological solutions by means of improved management or what is now sometimes called "ecological modernization." Fundamental issues of democracy are ignored from the outset, as is the nature of world capitalism, including its built-in dynamic of accumulation. In this one-sided view, only productive forces matter (along with human population trends); social organization and social relations do not.

The argument of *The Vulnerable Planet*, in contrast, is that environmental history is not simply a product of technological or demographic change; rather, it is inseparable from, even subordinate to, the way we organize our social relations. The environmental crisis is a crisis of society in the fullest sense. It signals the fact that a one-sided development of human productive powers without a commensurate change in the social relations by which we govern society spells social and ecological disaster. What is needed under present circumstances is an acceleration of history once again in the social realm, however terrifying that may be from a Burckhardtian viewpoint or a contemporary Worldwatch perspective. At issue is the possibility of a radical transformation of society: the creation of a society of equals dedicated to social justice and environmental sustainability.

From the standpoint of the hundreds of millions of years of geological time, the earth is of course immune to anything that we can do. It will recover from our greatest acts of malfeasance. Its "vulnerability" exists only in our limited, "parochial" (some would say "anthropocentric") outlook, which identifies the planet with our own species and those species with which we cohabit the earth. What is really at issue is not the survival of the earth as such but the integrity of the biosphere as we know it. It is "our" planet earth and not "the" planet earth which is in trouble.

What we must guard against the most as we endeavor to address our ecological crisis is the notion that history has in some way reached its end or that there are no genuine alternatives. Such beliefs of received ideology, if followed, would guarantee a barbaric, even apocalyptic, outcome. The real future of humanity, as distinct from this non-future, depends on the nature of our social and environmental movements, and ultimately on our willingness to reinvent human history and our social and ecological relations of production. There are no guarantees in this process. Historical developments are enormously contingent. The only certainty is the reality of the struggle itself: the true realm of human freedom.[5]

*Eugene, Oregon*
*Summer 1999*

# NOTES

## PREFACE

1. Raymond Williams, *The Country and the City* (New York: Oxford University Press, 1973), p. 302.
2. Marsh quoted in David Lowenthal, "Introduction," in George Perkins Marsh, *Man and Nature* (Cambridge, MA: Harvard University Press, 1965), p. xvii.

## 1: THE ECOLOGICAL CRISIS

1. Lester R. Brown et al., "World Without End," *Natural History* (May 1990): 89, and *State of the World 1992* (London: Earthscan, 1992), pp. 3-8.
2. Braudel quoted in Immanuel Wallerstein, *The Modern World System I* (New York: Academic Press, 1974), p. 128.
3. Donald Worster, ed., *The Ends of the Earth* (New York: Cambridge University Press, 1988), pp. 103-7; B.L. Turner II et al., eds., *The Earth as Transformed by Human Action* (New York: Cambridge University Press, 1990), p. 91; Charles Darwin, *The Voyage of the Beagle* (Garden City, NY: Doubleday, 1962), pp. 433-34; Roxanne Dunbar Ortiz, "Aboriginal Peoples and Impe-

rialism in the Western Hemisphere," *Monthly Review* 44, no. 4 (September 1992): 2; W.W. Rostow, *The World Economy* (Austin, TX: University of Texas Press, 1978), pp. 18-19.

4. Clive Ponting, *A Green History of the World: The Environment and the Collapse of Great Civilizations* (New York: St. Martin's Press, 1991), pp. 240-42; Worster, ed., *Ends of the Earth,* pp. 105-6.
5. Carlo M. Cipolla, *The Economic History of World Population* (Harmondsworth: Penguin Books, 1965), p. 89; Barry Commoner, *Making Peace with the Planet* (New York: The Free Press, 1992), pp. 160, 166; Turner et al., eds., *The Earth as Transformed,* pp. 42, 45-47; Colin McEvedy and Richard Jones, *Atlas of World Population* (Harmondsworth: Penguin Books, 1978), pp. 353-55; Scientific American, *Managing Planet Earth* (New York: W.H. Freeman, 1990), p. 61.
6. World Commission on Environment and Development, *Our Common Future* (New York: Oxford University Press, 1987), pp. 95-102.
7. Lewis Mumford, *Technics and Civilization* (New York: Harcourt Brace Jovanovich, 1963), pp. 107-11; Cipolla, *Economic History,* pp. 48-59; W.S. Jevons, *The Coal Question* (New York: Augustus M. Kelley, 1965), p. 2; Paul Harrison, *The Third Revolution* (New York: I.B. Tauris, 1992), pp. 34-36; Ponting, *Green History,* pp. 267-94; Donella Meadows, Dennis Meadows, and Jorgen Randers, *Beyond the Limits* (London: Earthscan, 1992), p. 66; World Commission on Environment and Development, *Our Common Future,* p. 169.
8. Paul Bairoch, "International Industrialization Levels from 1750 to 1980," *Journal of European Economic History* 11, no. 2 (Fall 1982): 269-99; World Commission on Environment and Development, *Our Common Future,* p. 206; Meadows et al., *Beyond the Limits,* pp. 5, 19, 37; Ponting, *Green History,* p. 325.
9. Paul Bairoch in Just Faaland, *Population and the World Economy in the 21st Century* (New York: St. Martin's Press, 1982), pp. 162-63; Robert Heilbroner, *Behind the Veil of Economics* (New York: W.W. Norton, 1988), pp. 54-55; Gerald Piel, *Only One World* (New York: W.H. Freeman, 1992), p. 11.
10. Turner et al., eds., *Earth as Transformed,* pp. 103-19; Ponting, *Green History,* pp. 295-314.
11. James O'Connor, "Uneven and Combined Development and Ecological Crisis," *Race and Class* 30, no. 3 (January-March 1989): 1-11; World Commission on Environment and Development, *Our Common Future,* pp. 235-41; Pete Hamill, "The Resurrection of Mexico City," *Audubon* 95, no. 1 (January-February 1993): 48; Frances Moore Lappé and Joseph Collins, *Food First* (New York: Ballantine Books, 1978), pp. 154-65.
12. Meadows et al., *Beyond the Limits,* pp. 44-46.
13. Ibid., pp. 47-56.
14. Ibid., pp. 57-66; Edward O. Wilson, *The Diversity of Life* (Cambridge, MA:

Harvard University Press, 1992), pp. 278-80, 346; Helen Caldicott, *If You Love This Planet* (New York: W.W. Norton, 1992), pp. 95-98; National Academy of Sciences, *One Earth, One Future* (Washington, DC: National Academy Press, 1990), pp. 123-25.

15. Meadows et al., *Beyond the Limits*, pp. 83, 89-91, 259.
16. Ibid., pp. 158-59; Caldicott, *If You Love This Planet*, pp. 17-22.
17. National Academy of Sciences, *One Earth, One Future*, pp. 67-71; Caldicott, *If You Love This Planet*, p. 24; Mustafa Tolba, *Saving Our Planet* (New York: Chapman and Hall, 1992), pp. 27-28; Jeremy Leggett, ed., *Global Warming* (New York: Oxford University Press, 1990), pp. 21-27, 158-59; Intergovernmental Panel on Climate Change, *Climate Change* (New York: Cambridge University Press, 1990), pp. xxii.
18. Intergovernmental Panel on Climate Change, *Climate Change*, p. xvi; Piel, *Only One World*, p. 10.
19. Paul Ehrlich and Anne Ehrlich, *Healing the Planet* (New York: Addison-Wesley, 1991), pp. 26-27; Jim MacNeill et al., *Beyond Interdependence* (New York: Oxford University Press, 1991), p. 8.
20. Peter M. Vitousek et al., "Human Appropriation of the Products of Photosynthesis," *Bioscience* (June 1986): 368; Meadows et al., *Beyond the Limits*, pp. 65-66.
21. Turner et al., eds., *Earth as Transformed*, pp. 1, 245.
22. Ibid., p. 1
23. Meadows et al., *Beyond the Limits*, pp. 97-103.
24. Harrison, *Third Revolution*, pp. 258, 345-46 (as Harrison notes, "Although *direct* energy use probably is not so skewed as income, when *indirect* use—via the additional products and services bought—is included, the assumption is probably a fair one."); Barry Commoner, "The Environmental Costs of Economic Growth," in Sam H. Schurr, ed., *Energy, Economic Growth and the Environment* (Baltimore: Johns Hopkins University Press, 1972), pp. 41-46.
25. See Meadows et al., *Beyond the Limits*, p. 100.
26. Paul M. Sweezy, "Capitalism and the Environment," *Monthly Review* 41, no. 2 (June 1989): 7.
27. Worster, ed., *Ends of the Earth*, p. 11.
28. Commoner, *Making Peace*, p. ix.

## 2: ECOLOGICAL CONDITIONS BEFORE THE INDUSTRIAL REVOLUTION

1. Yrjö Haila and Richard Levins, *Humanity and Nature* (London: Pluto Press, 1992), p. 190.
2. Ibid., pp. 192, 201.
3. J.D. Gould, *Economic Growth in History* (London: Methuen, 1972), p. 9. For

a critique, see Cipolla, *Before the Industrial Revolution* (New York: Norton, 1980), pp. 135-40.

4. Raymond Williams, *Resources of Hope* (London: Verso, 1989), p. 212.
5. Ponting, *Green History*, pp. 68-88.
6. Samir Amin, *Class and Nation* (New York: Monthly Review Press, 1980), pp. 46-70; L.S. Stavrianos, *Lifelines from Our Past* (New York: Pantheon, 1989), pp. 45-86; Eric Wolf, *Europe and the People Without History* (Berkeley: University of California Press, 1982), pp. 79-83.
7. Ponting, *Green History*, pp. 68-73; Daniel Hillel, *Out of the Earth* (Berkeley: University of California Press, 1991), pp. 78-87.
8. Ponting, *Green History*, pp. 73-83; Edward Goldsmith, *The Great U-Turn* (Hartland, Devon: Green Books, 1988), pp. 3-29; Tom Dale and Vernon Carter, *Topsoil and Civilization* (Norman, OK: University of Oklahoma Press, 1955), pp. 109-55, 196-97; T.P. Culbert, ed., *The Classic Maya Collapse* (Albuquerque: University of New Mexico Press, 1975), pp. 362-65.
9. Williams, *Country and the City*, pp. 37-38.
10. Kirkpatrick Sale, *The Conquest of Paradise* (New York: Alfred A. Knopf, 1990), pp. 74-91; Fernand Braudel, *The Structures of Everyday Life* (New York: Harper and Row, 1979), pp. 73-78.
11. Braudel, *Structures of Everyday Life*, pp. 70-71, 124, 158-71; Ponting, *Green History*, pp. 88-116.
12. Cipolla, *Before the Industrial Revolution*, p. 234; Braudel, *Structures of Everyday Life*, pp. 158-71; Jack Weatherford, *Indian Givers* (New York: Ballantine, 1988), pp. 59-78.
13. Montaigne and Bacon cited in J.H. Elliott, *The Old World and the New, 1492-1650* (New York: Cambridge University Press, 1970), pp. 87, 102-3.
14. Bacon, *The Great Instauration and New Atlantis* (Arlington Heights, IL: Harlan Davidson, 1980), pp. viii, 21, 31; Bacon quoted in Carolyn Merchant, *Radical Ecology* (New York: Routledge, 1992), p. 46; Carolyn Merchant, *The Death of Nature* (New York: Harper and Row, 1980), pp. 172-77.
15. Buffon cited in Worster, ed., *Ends of the Earth*, pp. 6-7.
16. Keith Thomas, *Man and the Natural World* (New York: Pantheon, 1983), pp. 274-75; Ponting, *Green History*, pp. 163-65, 177-85; Sale, *Conquest of Paradise*, pp. 261-62, 290; Carolyn Merchant, *Ecological Revolutions* (Chapel Hill: The University of North Carolina Press, 1989), pp. 43, 65-66; Peter Matthieson, *Wildlife in America* (New York: Viking, 1987), p. 81.
17. Ponting, *Green History*, pp. 195-96, 206-7.
18. Ibid.; Basil Davidson, "Columbus," *Race and Class* 33, no. 3 (1992): 17-25.
19. Wolf, *Europe and the People Without History*, pp. 195-96.
20. Jamaica Kincaid, "Foreword" to Guy Endore, *Babouk* (New York: Monthly Review Press, 1991), pp. vii-viii.
21. *Newsweek* (Columbus Special Issue), Fall/Winter 1991, p. 71; Ralph Davis,

*The Rise of the Atlantic Economies* (Ithaca, NY: Cornell University Press, 1973), p. 253.

22. Eduardo Galeano, *Open Veins of Latin America* (New York: Monthly Review Press, 1973), pp. 74-75.
23. Ibid.; Raynal quoted in Braudel, *Structures of Everyday Life,* p. 226.
24. Davis, *Rise of the Atlantic Economies,* p. 251; Braudel, *Structures of Everyday Life,* p. 224; Wallerstein, *Modern World System II,* p. 51; Stavrianos, *Global Rift,* pp. 96-97; Eric Williams, *Capitalism and Slavery* (New York: Capricorn Books, 1944), pp. 30-84.
25. Davis, *Rise of the Atlantic Economies,* p. 265; Ponting, *Green History,* p. 207; Avery Odelle Craven, *Soil Exhaustion as a Factor in the Agricultural History of Virginia and Maryland, 1606-1860* (Gloucester, MA: Peter Smith, 1965), p. 162; Sale, *Conquest of Paradise,* p. 291.
26. Anthony F.C. Wallace, *Death and Rebirth of the Seneca* (New York: Vintage, 1969), pp. 114-15; Francis Jennings, "The Indians' Revolution," in Alfred F. Young, ed., *The American Revolution* (Dekalb, IL: Northern Illinois University Press, 1976), pp. 333-35; George Washington, *Writings,* vol. II (New York: G.P. Putnam's Sons, 1889), pp. 220-22; Cecil B. Currey, *The Road to Revolution* (Garden City, NY: Doubleday, 1968), pp. 128-29.
27. Jesuit priest quoted in Howard Zinn, *A People's History of the United States* (New York: Harper and Row, 1980), pp. 19-20; Wallace, *Death and Rebirth,* pp. 111-14; Weatherford, *Indian Givers,* pp. 135-39.
28. W.J. Eccles, "The Fur Trade and Eighteenth Century Imperialism," in Alan Karras and J.R. McNeil, eds., *Atlantic American Societies* (New York: Routledge, 1992), pp. 212-34; Barbara Graymont, *The Iroquois in the Revolution* (Syracuse, NY: Syracuse University Press, 1976), pp. 104-28; Washington ("beasts of prey") quoted in Richard Drinnon, *Facing West* (Minneapolis: University of Minnesota Press, 1980), p. 65.
29. George Washington, "Instructions to General Sullivan, May 31, 1779," in John Sullivan, *Letters and Papers of Major General John Sullivan,* vol. 3 (Concord, NH: New Hampshire Historical Society, 1939), pp. 48-53.
30. Sullivan, *Letters and Papers,* vol. 3, pp. 134-37; William Stone, *The Life of Joseph Brant* (New York: Alexander V. Blake, 1838), pp. 34-42; Wallace, *Death and Rebirth,* pp. 141-44; Graymont, *Iroquois in the American Revolution,* pp. 213-19.
31. Wallace, *Death and Rebirth,* pp. 150, 184.
32. Merchant, *Ecological Revolutions,* pp. 186-88, 196-97; William Cronon, *Changes in the Land* (New York: Farrar, Straus & Giroux, 1983), pp. 43-48.

## 3: THE ENVIRONMENT AT THE TIME OF THE INDUSTRIAL REVOLUTION

1. Williams, *Country and City,* pp. 1-2, 60-61, 96-107, 279-88; Robert Brenner

in *The Brenner Debate* (Cambridge: Cambridge University Press, 1985), pp. 48-54, 213-17, 236-42, 324-25; J.V. Beckett, *The Agricultural Revolution* (Cambridge, MA: Basil Blackwell, 1990), pp. 1-10, 35, 60-61; C.P. Hill, *British Economic and Social History* (London: Edward Arnold, 1985), pp. 12-22.

2. Blake, *The Poems of William Blake* (London: Longman, 1971), p. 489; Hans-Magnus Enzensberger, "A Critique of Political Ecology," *New Left Review* 84 (March-April 1974): 9-10; Williams, *Resources of Hope*, pp. 211-12.
3. Herman Daly, *Steady-State Economics* (Washington: Island Press, 1991), p. 153.
4. E.J. Hobsbawm, *Industry and Empire* (Harmondsworth: Penguin Books, 1969), p. 34.
5. de Tocqueville quoted in E.J. Hobsbawm, *The Age of Revolution* (New York: Mentor, 1962), p. 44.
6. Ibid., pp. 51-54.
7. Adam Smith, *The Wealth of Nations* (New York: Modern Library, 1937), p. 17.
8. An unnamed visitor to England quoted in Hobsbawm, *Industry and Empire*, p. 56.
9. Charles Dickens, *Hard Times* (New York: W.W. Norton, 1966), pp. 17, 48.
10. Leo Huberman, *Man's Worldly Goods* (New York: Monthly Review Press, 1936), p. 186.
11. Paul Mantoux, *Industrial Revolution in the Eighteenth Century* (Chicago: University of Chicago Press, 1983), pp. 410-11; Lewis Mumford, *The City in History* (New York: Harcourt, Brace and World, 1961), p. 472. On the Babbage principle, see Harry Braverman, *Labor and Monopoly Capital* (New York: Monthly Review Press, 1974), pp. 79-82.
12. Engels, *The Condition of the Working Class in England* (Chicago: Academy Chicago, 1984), pp. 79-84.
13. Ibid., pp. 93-94.
14. Howard Waitzkin, *The Second Sickness* (New York: The Free Press, 1983), pp. 66-71; Engels, *Condition of the Working Class*, pp. 126-238 passim.
15. Mumford, *The Culture of Cities* (New York: Harcourt Brace and Co., 1938), pp. 162-64, 191-95; *City in History*, pp. 466-67; and *Technics and Civilization*, p. 158; Hobsbawm, *Industry and Empire*, pp. 86-87.
16. English M.P. and publicist Jerlinger Symons (1809-1860) cited in Engels, *Condition of the Working Class*, p. 167.
17. Huberman, *Man's Worldly Goods*, p. 180; Cipolla, *Economic History*, p. 101.
18. Thomas Malthus, "A Summary View on the Principle of Population," in D.V. Glass, ed., *Introduction to Malthus* (New York: John Wiley and Sons, 1953), pp. 123, 138; E.K. Hunt, *History of Economic Thought* (Belmont, CA: Wadsworth Publishing, 1979), pp. 65-66.
19. Thomas Malthus, *An Essay on the Principle of Population*, in Donald Winch,

ed. (Cambridge: Cambridge University Press), pp. 23, 29; Thomas Malthus, *Population: The First Essay* (Ann Arbor: University of Michigan Press), pp. 5, 30.

20. Raymond G. Cowherd, *Political Economists and the English Poor Laws* (Athens: Ohio University Press, 1967), pp. 161-64.
21. Malthus, *Population: The First Essay*, p. 29; Juan Martinez-Alier, *Ecological Economics* (Oxford: Basil Blackwell, 1987), p. 99.
22. Malthus, *Population: The First Essay*, p. 8.
23. Malthus, "Summary View," pp. 139-43; Harold Barnett and Chandler Morse, *Scarcity and Growth* (Baltimore: Johns Hopkins University Press, 1963), pp. 59-64.
24. Karl Marx, *Capital*, vol. I (New York: Vintage, 1976), pp. 783-84; Karl Marx, *Grundrisse* (New York: Vintage, 1973), pp. 606-8.
25. Karl Marx, *Wage-Labour and Capital and Value, Price and Profit*, pp. 33, 45, 58; Marx, *Capital*, vol. I, pp. 798-99.
26. *Marx, Capital*, vol. I, pp. 637-38.
27. Daniel Hillel, *Out of the Earth* (Berkeley: University of California Press, 1991), pp. 129-32, 292-93; Ponting, *Green History*, p. 199.
28. Karl Marx, *Capital*, vol. III (New York: Vintage, 1981), pp. 216, 949-50; Ronald Meek, "Introduction," in Marx and Engels, *Malthus* (New York: International Publishers, 1954), pp. 13-14, 28-31; Michael Perelman, *Marx's Crises Theory* (New York: Praeger, 1987), pp. 31-42.
29. Karl Marx, *Capital*, vol. II (New York: Vintage, 1978), pp. 321-22.
30. Frederick Engels, *The Dialectics of Nature* (New York: International Publishers, 1940), pp. 291-92; Marx, *Capital*, vol. I, pp. 134, 510, 638, 648-49; Perelman, *Marx's Crises Theory*, pp. 40-47.
31. Henry David Thoreau, *Walden and Other Writings* (New York: Bantam Books, 1962), pp. 124-25; Wordsworth, "On the Projected Kendal and Windemere Railway," in Wordsworth, *The Poems*, vol. II (Harmondsworth: Penguin, 1977), p. 889.
32. Thoreau, *Walden and Other Writings*, pp. 190-96.
33. John Ruskin, *Works*, vol. XVII, ed. E.T. Cook (New York: Longmans, Green and Co., 1905), pp. 87-89, and vol. XXVII (New York: Longmans, Green, 1907), p. 122; Raymond Williams, *Culture and Society, 1780-1950* (New York: Columbia University Press, 1983), pp. 142-43.
34. William Morris, *News from Nowhere and Selected Writings and Designs* (Harmondsworth: Penguin, 1962), pp. 85, 121-22.
35. Morris, "The Manifesto of the Socialist League," in E.P. Thompson, *William Morris* (New York: Pantheon, 1977), p. 734; Morris, *News from Nowhere*, pp. 121, 174-77; Williams, *Culture and Society*, p. 154.
36. Morris, *News from Nowhere*, pp. 131, 175.
37. Ibid., pp. 307-8.

## 4: EXPANSION AND CONSERVATION

1. Barnett and Morse, *Scarcity and Growth*, pp. 73-74; Karl Polanyi, *The Great Transformation* (Boston: Beacon Press, 1944), p. 178.
2. Richard B. DuBoff, *Accumulation and Power* (Armonk, NY: M.E. Sharpe, 1989), pp. 20, 32.
3. Frederick Jackson Turner, *The Frontier in American History* (New York: Henry Holt and Company, 1921), pp. 1, 11; William Cronon, *Nature's Metropolis* (New York: W.W. Norton, 1991), p. 31.
4. Turner, *Frontier in American History*, pp. 9, 15, 18-19, 213, 219.
5. Thorstein Veblen, *Absentee Ownership* (New York: Augustus M. Kelley, 1923), pp. 122-26, 168-71, 186-91.
6. Marsh, *Man and Nature*, pp. 42-43; Worster, ed., *Ends of the Earth*, pp. 7-8; Lewis Mumford, *The Brown Decades* (New York: Dover, 1971), p. 35.
7. Marsh, *Man and Nature*, pp. ix, 35-36, 42-43.
8. See Worster, ed., *Ends of the Earth*, pp. 8-14.
9. Lawrence Buell, "American Pastoral Ideology Reappraised," in Thoreau, *Walden and Resistance to Civil Government*, ed. William Rossi (New York: W.W. Norton, 1992), pp. 473-74; *Audubon Magazine* 1, no. 1 (February 1887): 9, 13-14.
10. Ponting, *Green History*, pp. 167-68; Cronon, *Nature's Metropolis*, pp. 213-18; E. Franklin Frazier, *Race and Culture Contacts in the Modern World* (Boston: Beacon, 1957), pp. 61-62; Richard Irving Dodge, *The Plains of the Great West* (New York: G.P. Putnam, 1877), pp. 119-47.
11. Samuel P. Hays, *Conservation and the Gospel of Efficiency* (Cambridge, MA: Harvard University Press, 1959), p. 5.
12. Pinchot quoted in ibid., pp. 41-42, 191.
13. Quoted in J. Leonard Bates, "Fulfilling American Democracy," *Mississippi Valley Historical Review* 44, no. 1 (June 1957): 41-42.
14. Theodore Roosevelt, *Works*, vol. 16, ed. Hermann Hagedorn (New York: Charles Scribner's Sons, 1926), pp. 101, 105.
15. Aldo Leopold, *A Sand County Almanac* (New York: Oxford University Press, 1949), pp. 204, 207-14; Hays, *Conservation and the Gospel of Efficiency*, pp. 192-95; Roderick Nash, *Wilderness and the American Mind* (New Haven: Yale University Press, 1967), pp. 161-81.
16. Robert Marshall, *The People's Forests* (New York: Harrison Smith and Robert Haas, 1933), pp. 123, 148-49, 209-19; Robert Gottlieb, *Forcing the Spring* (Washington, DC: Island Press, 1993), pp. 8, 15-19; James M. Glover, *A Wilderness Original* (Seattle: The Mountaineers, 1986).
17. Mumford, *City in History*, pp. 467-68; Sidney Webb, *Socialism in England* (New York: Charles Scribner's Sons, 1901), pp. 101-02.
18. Mumford, *City in History*, p. 476; Francis Sheppard, *London 1808-1870* (London: Secker and Warburg, 1971), pp. 247-96.
19. Martin Melosi, ed., *Pollution and Reform in American Cities, 1870-1930*

(Austin: University of Texas Press, 1980), pp. 173-98; Gottlieb, *Forcing the Spring*, pp. 47-51, 59-63, 216-17.

20. Mumford, *Culture of Cities*, pp. 392-401; *City in History*, pp. 514-24.
21. Franklin Folsom, *Impatient Armies of the Poor* (Niwot, CO: University Press of Colorado, 1991), pp. 83-107.
22. Edward Bellamy, *Looking Backward* (New York: New American Library, 1960), pp. 213, 43.
23. Albert Fein, *Frederick Law Olmsted and the American Environmental Tradition* (New York: George Braziller, 1972), pp. 57-61; Mumford, *Brown Decades*, p. 40; Bellamy, *Looking Backward*, pp. x, 165.
24. Upton Sinclair, *The Jungle: The Lost First Edition* (Memphis, TN: Peachtree Publishers, 1988), p. 83.
25. Cronon, *Nature's Metropolis*, p. 259; Sinclair, *The Jungle*, p. 283.
26. Sinclair, *The Jungle*, pp. 28-29.

## 5: IMPERIALISM AND ECOLOGY

1. Wolf, *Europe and the People Without History*, pp. 232-39.
2. Ibid., pp. 232-51; Stavrianos, *Global Rift*, pp. 242-46.
3. Wolf, *Europe and the People Without History*, p. 248; Stavrianos, *Global Rift*, pp. 315-19; Jack Beeching, *The Chinese Opium Wars* (New York: Harcourt Brace Jovanovich, 1975); Hobsbawm, *Age of Revolution*, p. 355.
4. Hobsbawm, *The Age of Empire* (New York: Pantheon, 1987), pp. 56-59.
5. Rhodes quoted in Sarah Gertrude Millin, *Rhodes* (London: Chatto and Windhurst, 1952), p. 138; Belloc cited in Hobsbawm, *Age of Empire*, p. 20.
6. Cited in *The Ecologist* 22, no. 4 (July-August 1992): 134.
7. Turner, *Frontier in American History*, p. 219.
8. Quoted in Zinn, *People's History of the United States*, p. 303.
9. Andre Gunder Frank, *Capitalism and Underdevelopment in Latin America* (New York: Monthly Review Press, 1969).
10. Daniel Headrick, *Tentacles of Progress* (New York: Oxford University Press, 1988), pp. 55-91.
11. Ibid., p. 147.
12. Ibid., pp. 165-67.
13. Josué de Castro, *The Geography of Hunger* (Boston: Little, Brown and Co., 1952), pp. 192-93, 213-15, 221-22, 284-65.
14. Headrick, *Tentacles of Progress*, p. 209.
15. Lucille Brockway, *Science and Colonial Expansion* (New York: Academic Press, 1979), p. 190; Headrick, *Tentacles of Progress*, pp. 211-15.
16. Headrick, *Tentacles of Progress*, pp. 243-50; Lappé and *Collins, Food First*, p. 157; Cary Fowler and Pat Mooney, *Shattering* (Tucson: University of Arizona Press, 1990).

17. Lappé and Collins, *Food First*, pp. 156-60; Fowler and Mooney, *Shattering*, pp. xii-xiv; Kloppenburg, *First the Seed* (New York: Cambridge University Press, 1988), p. 161.
18. Onate quoted in Vandana Shiva, *The Violence of the Green Revolution* (London: Zed Books, 1991), p. 44; Fowler and Mooney, *Shattering*, p. 183; Larry Everest, *Behind the Poison* (Chicago: Banner Press, 1985).
19. Harry Magdoff, "Perestroika and the Future of Socialism," Parts I and II, *Monthly Review* 41, nos. 11 and 12 (March/April 1990); Greg McLauchlan, "The End of the Cold War as a Social Process," in Louis Kriesberg and David R. Segal, ed., *Research in Social Movements, Conflicts and Change*, vol. 14 (Greenwich, Conn.: JAI Press, 1992), p. 61. On the early history of Soviet ecology, see Douglas R. Weiner, *Models of Nature* (Bloomington: Indiana University Press, 1988).
20. Aganbegyan quoted in Magdoff, "Perestroika," Part II, p. 5.
21. Ibid.
22. Ibid.; Moshe Lewin, *Political Undercurrents in Soviet Economic Debates* (Princeton, NJ: Princeton University Press, 1974), p. 133.
23. Magdoff, "Perestroika," Part I, pp. 12-13; Part II, pp. 6-7.
24. Ibid., Part I, pp. 12-13.
25. Yablokov quoted in Murray Feshbach and Alfred Friendly, Jr., *Ecocide in the USSR* (New York: Basic Books, 1992), p. 9.
26. Ibid., pp. 2, 60-67.
27. Ibid., pp. 57-58, 93-97, 124; Brown et al., *The State of the World 1991*, p. 96; Commoner, *Making Peace with the Planet*, p. 220.
28. Feshbach and Friendly, *Ecocide in the USSR*, p. 12; *Earthwatch* 12, no. 5 (July/August 1993): 5; Tim Deere-Jones, "Back to the Land," *The Ecologist* 21, no. 1 (January/February 1991): 19.
29. Gabriel Kolko, *Anatomy of a War* (New York: Pantheon, 1985), p. 189; Stavrianos, *Global Rift*, p. 726.
30. Kolko, *Anatomy of a War*, pp. 144-45, 238-39; Paul Frederick Cecil, *Herbicidal Warfare* (New York: Praeger, 1986), pp. 23-35, 38, 54, 109.
31. Arthur H. Westing, ed., *Herbicides in War* (Philadelphia: Taylor and Francis, 1984), p. 22.
32. Harry Magdoff, "Globalization—To What End?" in Ralph Miliband and Leo Panitch, eds., *The Socialist Register 1992* (New York: Monthly Review Press, 1992), p. 48; World Bank, *World Development Report* (New York: Oxford University Press, 1991), p. 14.
33. Budget of the United States Government, Fiscal Year 1992 (Washington, DC: U.S. Government Printing Office, 1991), Part Seven 68-70; Paul M. Sweezy, "U.S. Imperialism in the 1990s," *Monthly Review* 41, no. 5 (October 1989): 10.
34. Daniel Faber, *Environment Under Fire* (New York: Monthly Review Press, 1993), pp. 154-80.

35. Ramsey Clark, *The Fire This Time* (New York: Thunders Mouth Press, 1992), pp. 94-108; T.M. Hawley, *Against the Fires of Hell* (New York: Harcourt Brace Jovanovich, 1992), p. 142.
36. Johan Holmberg, ed., *Making Development Sustainable* (Washington, DC: Island Press, 1992), p. 330; World Bank, *World Development Report 1993* (New York: Oxford University Press, 1993), p. 199.
37. Holmberg, ed., *Making Development Sustainable,* pp. 335-36.
38. Ibid., p. 323; Germaine Greer, *Sex and Destiny* (London: Picador, 1985), p. 409.

## 6: THE VULNERABLE PLANET

1. *Statistical Abstract of the United States,* 1986 (Washington, DC: U.S. Government Printing Office, 1986), p. 524.
2. Braverman, *Labor and Monopoly Capital,* p. 156; also Marx, *Capital,* vol. I, p. 1035.
3. Richard Sasuly, *IG Farben* (New York: Boni and Gaer, 1947), p. 19.
4. Ibid., pp. 163-66; Robert Bruce Lindsay, *The Role of Science in Civilization* (New York: Harper and Row, 1963), pp. 214-19.
5. Commoner, *Closing Circle,* pp. 128-30.
6. Braverman, *Labor and Monopoly Capital,* pp. 166-67.
7. Ibid., pp. 112-21; Frederick Winslow Taylor, *The Principles of Scientific Management* (New York: W.W. Norton, 1947).
8. Elliott A. Norse, *Ancient Forests of the Pacific Northwest* (Washington, DC: Island Press, 1990), pp. 152-60.
9. Paul Baran and Paul Sweezy, *Monopoly Capital* (New York: Monthly Review Press, 1966), p. 132.
10. Matthew Edel, *Economics and the Environment* (Engelwood Cliffs, NJ: Prentice-Hall, 1973), p. 136; data from Commoner, *Closing Circle,* pp. 140-41.
11. Commoner, *Closing Circle,* pp. 138, 175.
12. Schurr, ed., *Energy, Economic Growth, and the Environment,* pp. 44-62.
13. Bradford Snell, "American Ground Transport," in U.S. Senate, Committee on the Judiciary, *Industrial Reorganization Act,* Hearings Before the Subcommittee on Antitrust and Monopoly, 93rd Congress, 2nd Session, Part 4a (Washington, DC: U.S. Government Printing Office, 1974), pp. A-26, A-47; Glenn Yago, "Corporate Power and Urban Transportation," in Maurice Zeitlin, *Classes, Class Conflict, and the State* (Cambridge, MA: Winthrop Publishers, 1980), pp. 296-323; Motor Vehicles Manufacturers Association of the United States, *MVMA Facts and Figures* (Detroit, MI: MVMA, 1990), p. 36.
14. Rolf Edberg and Alexi Yablokov, *Tomorrow Will Be Too Late* (Tucson: University of Arizona Press, 1991), pp. 92, 101.

15. Barry Commoner, "Preface," in Michael Perelman, *Farming for Profit in a Hungry World* (Montclair, NJ: Allanheld, Osmun & Co., 1977), p. vii.
16. R.C. Lewontin and Jean-Pierre Berlan, "Technology, Research, and the Penetration of Capital," *Monthly Review* 38, no. 3 (July-August 1986), pp. 21, 26-27.
17. Jean-Pierre Berlan and R.C. Lewontin, "The Political Economy of Hybrid Corn," *Monthly Review* 38, no. 3 (July-August 1986): 35-47; R.C. Lewontin, *Biology as Ideology* (New York: HarperCollins, 1991), pp. 54-57.
18. Commoner, *Closing Circle*, pp. 29-42; Edberg and Yablokov, *Tomorrow Will Be Too Late*, p. 89; Haila and Levins, *Humanity and Nature*, pp. 5-6. Although Commoner refers to the fourth law as "there's no such thing as a free lunch," the Russian scientist Yablokov has translated this more generally as "nothing comes from nothing."
19. Commoner, *Closing Circle*, pp. 37-41; and *Making Peace*, pp. 11-13. Commoner's third law should not be taken too literally. As Haila and Levins write, "The conception that 'nature knows best' is relativized by the contingency of evolution." Haila and Levins, *Humanity and Nature*, p.6.
20. Ibid., pp. 14-15; Herman E. Daly and Kenneth Townsend, eds., *Valuing the Earth* (Cambridge, MA: MIT Press, 1993), pp. 69-73.
21. Donald Worster, *The Wealth of Nature* (New York: Oxford University Press, 1993), pp. 58-59.
22. Vandana Shiva, *Staying Alive* (London: Zed Books, 1989), pp. 23-24, 186.
23. Haila and Levins, *Humanity and Nature*, p. 201.
24. Nicholas Georgescu-Roegen, *The Entropy Law and the Economic Process* (Cambridge: Harvard University Press, 1971), p. 2.
25. Commoner, *Making Peace*, pp. 10-11.
26. Haila and Levins, *Humanity and Nature*, p. 160.
27. Georgescu-Roegen, *Entropy Law*, p. 2; K. William Kapp, *The Social Costs of Private Enterprise* (Cambridge, Mass.: Harvard University Press, 1971), p. 231.
28. Chandler Morse, "Environment, Economics and Socialism," Monthly Review 30, no. 11 (April 1979): 12; Commoner, *Making Peace*, pp. 82-83; and *The Poverty of Power* (New York: Alfred A. Knopf, 1976), p. 194.
29. Ford and DeLorean quoted in Commoner, *Making Peace*, pp. 80-81.
30. Robert Heilbroner, *An Inquiry into the Human Prospect* (New York: W.W. Norton, 1980), p. 100.

## 7: THE SOCIALIZATION OF NATURE

1. Commoner, *Making Peace*, pp. 20-28, 38, 41; Eric Mann, *L.A's Lethal Air* (Los Angeles: Labor/Community Strategy Center, 1991), p. 39.
2. Shirley Briggs, "Silent Spring: A View from 1990," *The Ecologist* 20, no. 2 (March/April 1990), pp. 54-55.

3. Gary Cohen and John O'Connor, eds., *Fighting Toxics* (Washington, DC: Island Press, 1990), pp. 12-15; 271-73; Commoner, *Making Peace*, pp. 31-32.
4. Kathryn A. Kohm, ed., *Balancing on the Brink of Extinction* (Washington, DC: Island Press, 1991), pp. 77-85.
5. Brown et al., *The State of the World 1992*, p. 174; Edward Goldsmith, *The Way* (London: Rider, 1992), p. xi.
6. Marilynne Robinson, *Mother Country* (New York: Ballantine, 1989), pp. 1-8.
7. Commoner, *Making Peace*, pp. 41-44; Lester Brown et al., *Vital Signs 1992* (New York: W.W. Norton, 1992), pp. 70-71.
8. Daniel Faber and James O'Connor, "The Struggle for Nature," *Capitalism, Nature, Socialism* 2 (Summer 1989): 23-25; J. B. Foster, "Limits of Environmentalism," in *Capitalism, Nature, Socialism* 4, no. 1 (March 1993): 22-29.
9. Brown et al., *State of the World 1992*, p. 174.
10. Ibid., pp. 189-99; Robert Goodland, Herman Daly, and Salah El Serafy, eds., *Population, Technology and Lifestyle* (Washington, DC: Island Press, 1992).
11. Enzensberger, "A Critique of Political Ecology," p. 26.
12. World Commission on Environment and Development, *Our Common Future*, pp. 43, 52, 89; Herman Daly, "Sustainable Growth" in Daly and Townsend, eds., *Valuing the Earth*, pp. 267-73; W.M. Adams, *Green Development* (New York: Routledge, 1990), pp. 57-62.
13. Meadows et al., *Beyond the Limits*, pp. xv, 60-61, 209; Sweezy, "Capitalism and the Environment," p. 6.
14. World Commission on Environment and Development, *Our Common Future*, p. 50; Magdoff, "Globalization," p. 48; David Barkin, "Environmental Degradation in Mexico," *Monthly Review* 45, no. 3 (July-August 1993): 73-74.
15. István Mészáros, *The Necessity of Social Control* (London: Merlin Press, 1971), pp. 63-64; Magdoff, "Are There Lessons to Be Learned?" *Monthly Review* 42, no. 9 (February 1991): 12-13.
16. Amartya Sen, "The Economics of Life and Death," *Scientific American* 268, no. 5 (May 1993): 44-45; Richard W. Franke and Barbara H. Chasin, "Kerala State, India," *Monthly Review* 42, no. 8 (January 1991): 1-23; Vicente Navarro, "Has Socialism Failed?" *Science and Society* 57, no. 1 (Spring 1993): 15-21.
17. Penny Kemp et al., *Europe's Green Alternative* (London: Green Print, 1992), pp. 16-17.
18. Polanyi, *Great Transformation*, p. 132, emphasis added; Manfred Bienefeld, "Lessons of History and the Developing World," *Monthly Review* 41, no. 3 (July-August 1989): 14.
19. Lois Gibbs, "Foreword," in Nicholas Freudenberg, ed., *Not in Our Backyards!* (New York: Monthly Review Press, 1984), pp. 9-10.

20. Merchant, *Radical Ecology*, pp. 159-62; Carl Anthony et al., "A Place at the Table," *Sierra* 78, no. 3 (May-June 1993): 53.
21. Kirkpatrick Sale, *The Green Revolution* (New York: Hill and Wang, 1993), pp. 65-66, 76, 85; Faber and O'Connor, "The Struggle for Nature," pp. 32-33.
22. Gottlieb, *Forcing the Spring*, pp. 296-97, 301-2.
23. Merchant, *Radical Ecology*, pp. 164-67; Robert Bullard, ed., *Confronting Environmental Racism* (Boston: South End Press, 1993), p. 17; United Church of Christ, Commission for Racial Justice, *Toxic Wastes and Race in the United States* (New York: United Church of Christ, 1987), p. 23; Anthony et al., "Place at the Table," p. 51.
24. Anthony et al., "A Place at the Table," p. 58.
25. Southwest Organizing Project, Letter to Group of Ten environmental organizations, Albuquerque, New Mexico, March 16, 1992.
26. Merchant, *Radical Ecology*, pp. 200-2; Shobita Jain, "Women's Role in the Chipko Movement," in Sally Sontheimer, ed., *Women and the Environment* (New York: Monthly Review Press, 1991), pp. 163-78.
27. Shiva, *Staying Alive*, p. 76, quoted in Merchant, *Radical Ecology*, p. 201.
28. Merchant, *Radical Ecology*, pp. 224-25; Chico Mendes, *Fight for the Forest* (New York: Monthly Review Press, 1992).
29. Susanna Hecht and Alexander Cockburn, *The Fate of the Forest* (New York: HarperCollins, 1990), pp. 238-39.
30. Edberg and Yablokov, *Tomorrow Will Be Too Late*; A.V. Yablokov and S.A. Ostroumov, *Conservation of Living Nature and Resources* (New York: Springer-Verlag, 1991), p. 140.

## AFTERWORD

1. Richard Leakey and Roger Lewin, *The Sixth Extinction* (New York: Doubleday, 1995).
2. Jacob Burckhardt, *Reflections on History* (Indianapolis: Liberty Press, 1979), 213, 224.
3. Lester R. Brown et al., *The State of the World 1999* (New York: W. W. Norton, 1999), 4-11.
4. Ibid., 8, 11-12, 179; Lester R. Brown et al., *The State of the World 1996* (New York: W. W. Norton, 1996), 3-20.
5. Daniel Singer, *Whose Millennium?* (New York: Monthly Review Press, 1999).

# INDEX